The Homestead and Small Farm Legal Guide

Right to Farm

by David Lee Mundy

The Homestead and Small Farm Legal Guide: Right to Farm
© 2017
ISBN 978-1-7378845-8-3

About Handong Global University
This book was completed with assistance from Handong Global University on a leave for sabbatical to the states during which time the author worked alongside Farm-to-Consumer Legal Defense Fund (FTCLDF). Any opinions are the authors own.

About Farm to Consumer Legal Defense Fund
The Farm-to-Consumer Legal Defense Fund (FTCLDF) is a member based 501(c)(4) non-profit organization. FTCLDF protects the rights of farmers and consumers to engage in direct commerce; it protects the rights of farmers to sell the products of the farm and the rights of consumers to access the foods of their choice from the source of their choice.

Compilation
Any updates, corrections, or suggestions for additional sections in future editions should be submitted to mundy@handong.edu.

Legal Disclaimer
This guide is intended for educational and informational purposes only and is not intended to be nor should it be construed as either a legal opinion or as legal advice. Legal information was last checked on March 2017.

Table of Contents

Preface & Legal Disclaimer

Heather Retberg, a farm rights advocate in Maine, leaned across the dinner table. She dropped her voice to a conspiratorial whisper.

Finally, she was going to share the secret behind the overregulation of farming and cottage foods!

The answer surprised us.

It wasn't the illuminati, the Rothschilds, or Big Ag behind the draconian regulations.

Heather whispered, "It's the sink manufacturers!"

She leaned back, a smile played across her face. Her pal, Bonnie Preston, nodded along. It was obviously an old joke between the two of them.

Hard to believe this unassuming pair was the driving force behind local food ordinances in Maine. Last year, they almost passed an amendment to the state constitution.

Heather followed up on her sink-manufacturer joke with a real-life story of a storeowner who had been forced to move his sink again and again, and to purchase larger and larger sinks, all in order to comply with the latest regulations.

That type of story is all too familiar to our readers.

Heather might have been joking about the sink conspiracy. Nevertheless, small farmers, homesteaders, or cottage food producers might sometimes feel like there is a vast, governmental conspiracy aligned against them.

It's not a conspiracy. At least not a formal one.

There is, perhaps, an alignment of interests between risk-

averse bureaucrats and an established food industry, which results in regulations that suppress competition.[1]

There are also real concerns about food safety.

The question, then, is how to navigate the Serbonian bog of
rules and regulations?

Don't despair. There is hope and there is help available. But
as you'll soon discover, no one can do the work for you. You
have to educate yourself on the laws same as you do on crop
rotations, fertilizers, and other agricultural science.

Please allow us a small caveat.

Farmers are practical and might get frustrated by lawyerly,
weasel words. Lawyers speak and write with a frustrating lack
of certainty. Mr. Dickens had us dead-to-rights in *Great Expecta-
tions* with his character Mr. Jaggers. But as a farmer relies on past
patterns to predict the weather, lawyers can only do the same
with the law, which can be even more capricious and mercurial
than a spring storm.

You'll forgive me, then, and please understand that I am legally obligated to add the following disclaimer:

**This resource is not intended as legal advice. It is not
intended to, and cannot substitute for, sound legal
advice from a competent, licensed attorney
in your state.**

That may seem like good, common sense to you and to me,
but we are dealing with the same type of bureaucratic mentality
that takes children's medicine off the shelf because some parents
can't be bothered to read the label. It's a mad world.

---

[1] Nothing new about using the government to protect one's industry from competition. Businesses that can't compete have often been accused of lobbying Washington
to erect barriers of entry. Thomas DiLorenzo catalogues such case studies in his book
"How Capitalism Saved America: The Untold History of Our Country, from the Pil-
grims to the Present." There's also the cautionary tale of the epi-pen, a life-saving
medical device whose price soared based on corporate greed coupled with government protectionism: www.zerohedge.com/news/2016-08-27/epipen-scandal-worse-
you-think-what-you%E2%80%99re-not-being-told.

On a positive note, though you may not believe it, there is something positive to be said about regulations. The federal government and most states have completely discarded our common law heritage. However, by doing so, by adopting a civil law system, the government must *write everything down*. If it is not in writing, it is not law.

I'll never forget the look on one farmer's face as he caressed a mobile chicken coup. He loved that coop because it wasn't defined in the law books as farm equipment or as a building, which made it essentially unregulated!

Unfortunately, that's the exception.

The resources provided by FTCLDF are intended to help you, the farmer, navigate the web of laws, regulations, local ordinances, and zoning restrictions that comprise and complicate the field of agriculture and food production law. We hope to help keep your business moving.

Again, the following resource serves as a general overview of Right to Farm laws throughout the country. You should never assume, however, that the following information applies to your specific case without consulting an attorney in your home state.

Furthermore, though we've made efforts to ensure the accuracy of the information contained below, we can't guarantee it's correct. Laws and regulations can change at any time. Hence, you should take independent steps to ensure the legal information you have is up-to-date before relying on it in any way.

We have linked to URLs whenever possible purely for the convenience of the reader. We don't vouch for the contents of those sites and disclaim any liability for their content.

Finally, we have at times provided an opinion, findings, conclusions or recommendations beyond the text of the law itself. That represents the personal views of the author Any feedback is appreciated and may be sent to: mundy@handong.edu.

Right to Farm

Introduction

We all have a crazy neighbor who makes things awkward for the rest of us.

For instance, I have a friend whose neighbor sits in a lawn chair facing their house, staring at them for hours on end.

Then there are the neighborhood busybodies, like the neighbor who called our Texas grandma saying, "Viva, there's children climbing in your pecan trees!"

"I know," Grandma replied in her dry, sardonic tone. "They're my grandchildren!"

Then there's the city-slickers who complain about country life. One farmer got an earful from a passing motorist, complaining that his cows' udders look red and swollen. The farmer patiently explained to the urbanite that meant the cows were producing milk![2]

Problems between farmers and non-farming neighbors are so common that they have become fodder for internet memes. A funny sign making the rounds on social media reads:

> NOTICE: This property is a farm. Farms have animals. Animals make funny sounds, smell bad and have sex outdoors. Unless you can tolerate noise, odors and outdoor sex, don't buy property next to a farm!

For residential homeowners, bothersome neighbors can be annoying or even endearing at times.

---

[2] Tiffany Dowell, Student Note, "Daddy Won't Sell the Farm: Drafting Right To Farm Statutes To Protect Small Family Producers," 18 S.J. Agric. L. Rev. 127 (2009), citing Neil D. Hamilton, "A Livestock Producer's Legal Guide to: Nuisance, Land Use Control, and Environmental Law" 68 (1992).

Farmers, on the other hand, can have their livelihoods destroyed by such neighbors.

For this very reason, Right to Farm laws exist in all fifty states.

Right to Farm laws protect existing farm operations from becoming "nuisances" when non-agricultural uses invade the surrounding area.[3]

According to one legal encyclopedia,[4] these laws prevent neighbors, especially those who have moved into the neighborhood, from claiming that the sights, sounds, and smells of a farm, create a nuisance.[5]

Put simply, Right to Farm laws keep *existing* farms from being sued out of business by new, residential neighbors.

The Right to Farm laws of many states provide a one year time limit after which lawsuits may not be brought. Technically speaking, many courts have found this time limit to be a "statute of repose." One aspect of a statute of repose is that the time limit begins to run when the operation begins, not on the date the nuisance is "discovered" by the neighbors. For instance, a CAFO might start in January, but the neighbors might not discover obnoxious odors till June. In such instances, courts may start the clock running from January anyway.[6]

While we're dealing with technicalities, many Right to Farm laws are viewed as an "affirmative defense," which means if it is not raised before or during trial, it may not be raised for the first time on appeal.

Moving on.

---

[3] Neil D. Hamilton, "Right-To-Farm Laws Reconsidered: Ten Reasons Why Legislative Efforts To Resolve Agricultural Nuisances May Be Ineffective," 3 Drake J. Agric. L. 103, 103 (1998). Hamilton states Right to Farm laws represent a codification of the common law notion of "coming to the nuisance."
[4] 3 Am Jur 2d Agriculture § 8.
[5] Nuisance is defined as "Anything done by one which annoys or disturbs another in the free use, possession, or enjoyment of his property, or which renders its ordinary use or occupation physically uncomfortable." Ballentine's Law Dictionary 3d (LexisNexis 2010).
[6] *See Trinity River Auth. v. URS Consultants*, 889 S.W.2d 259, 261 (Tex. 1994).

Since Right to Farm typically protects existing farms, new farms or farms that spring up in urban areas may not receive protection.

While we are busting myths, let's be clear, Right to Farm does not mean you have the right to do whatever you want to on your property, especially if that property is zoned residential and has never been used for farming.

For instance, you don't have a "right" to raise chickens or grow vegetables in your urban backyard. Would that it were so. It ain't.

Right to Farm laws are intended to prevent "coming to the nuisance" claims.

Let's bust another myth while we're at it. You might think nuisance complaints are limited to large, containment feeding operations like commercial swine and poultry feedlots.

They aren't.

According to a study in New York, 25% of small dairy farms in the state had received a complaint in the past five years as had 33% of all small farms.[7]

Given those numbers, and how economically damaging a nuisance lawsuit can be to small farmers, it might seem that Right to Farm laws make good, common sense.

Yet Right to Farm laws are not without controversy. Critics argue that the American Legislative Executive Council (ALEC), the group behind Right to Farm laws, drafted the laws to protect its main constituents: industrial and confinement operations. It's no surprise then, that environmental and animal rights advocacy groups have come out against Right to Farm laws.

There is a history of bad blood between farming operations and animal rights groups. In fact, several states have passed "ag-

---

[7] Dowell, supra n. 3 citing a phone conversation with and an article by Lee Telega, "You Have A Right-to-Farm: Use It Wisely," SMALL FARM QUARTERLY, Jan. 10, 2005, at 6, available at www.smallfarms.cornell.edu/pages/quarterly/archive /winter05/winter05-06.pdf (surveying dairy farmers in New York state).

gag" laws that make it a criminal offense to make secret recordings in animal facilities.[8]

However, as to Right to Farm laws, there appear to be valid
points on both sides.

Right to Farm laws protect farmers from urban encroachment. However, those same laws may prevent neighbors, even
other neighboring farmers, from contesting the conversion of
traditional small farms into concentrated animal feeding operations (CAFOs).[9]

Right to Farm statutes were created in an era of smaller, fam-
ily-owned farms. However, some of those laws now give blanket
immunity to CAFOs and other large scale, commercial agricultural operations.

Also, because Right to Farm laws prevent neighbors from
bringing nuisance lawsuits, they might look to other avenues for
legal redress. A recent study indicated that living near CAFOs
may impair lung function.[10] These types of studies may point to
alternative tort options on the horizon.

Some courts have rejected "creative" legal theories. For instance, the courts in many jurisdictions have said that trespass
and nuisance actions are essentially the same. Characterizing the
case as a "trespass" cannot be used to get around a ban on nuisance liability.[11]

On the other hand, Right to Farm protection doesn't prevent
every nuisance lawsuit, especially where the farm substantially

---

[8] Animal Legal Defense Fund, "Taking Ag-gag to Court," available at aldf.org/cases-
campaigns/features/taking-ag-gag-to-court (last visited Oct. 25, 2016).
[9] David Bly, member of the house of representatives of the state of Minnesota published a well-reasoned op-ed on this issue: "Supporting farmers in Minnesota is a
bipartisan issue," *available at* davidbly.com/archive/supporting-farmers-in-minne-
sota-is-a-bipartisan-issue.
[10] www.ft.com/content/44588db4-705f-11e6-9ac1-1055824ca907.
[11] States including Texas, Oregon, California and Hawaii have held as much. Tiffany
Dowell Lashmet, "Legal: Understand your right to farm," CattleNetwork.com, Oct.
05, 2015, 5:00 am EDT, www.cattlenetwork.com/news/legal-understand-your-right-
farm.

changed its operations.

In addition, at least one state supreme court held that the Right to Farm law, by preventing neighbors from suing in nuisance, constituted an unconstitutional "taking" of the neighbors' land![12]

Perhaps the best compromise between the opposing factions is proposed by farmer Joel Salatin who runs Polyface Farms in Virginia. Salatin says farming should be a "sensuous relationship" with the animals and the earth. "If it smells bad," says Joel, "it's probably not good farming."

Many states, therefore, focus on education and on dispute resolution and mediation to smooth over disputes between farms and residential neighbors. Some states require sellers to provide notice to potential buyers of potential agricultural nuisances.

Regardless, Right to Farm laws are an important tool in the fight to defend the right of families to produce their own food for consumption and sale.

Right to Farm not only protects farmers from litigious neighbors, it can also provide a defense against overzealous zoning boards.

Take, for instance, a common problem farmers face in having a rich or otherwise influential neighbor move into their area who one day decides she no longer likes farming and does her best to hurt the farmer's operation even though the area is either zoned for agriculture or is zoned where agriculture is a permitted use. Even if there's a Right to Farm, what happens when the influential, rich neighbor seeks a restrictive zoning or health regulation?

---

[12] *Bormann v. Board of Sup'rs In and For Kossuth County*, 584 N.W.2d 309 (Iowa 1998).

Mighty Ducks

That's what happened to Kara Zaks an FTCLDF member in Massachusetts.[13]

After hatching ducklings in a classroom with her school children, Kara fell in love with the little critters and, as with many who catch the breeding bug, went on to care for some rare breeds, about twenty ducks in all.

According to Kara, her town's zoning and by-laws allowed for agricultural animals in any number except pigs.

She intentionally bought property in a community that had a right-to-farm in all zones of the town, even commercial. She dutifully complied with zoning ordinances including setbacks and provisions relating to coops and ponds. (We will get to those sorts of requirements in a later publication on zoning. Suffice it to say, Kara did her homework and complied with the local rules).

Her family was eating fresh eggs, they were conserving rare duck breeds, they were using duck manure for their garden. Years passed.

Then a neighbor objected.

Kara says the neighbor called the Board of Health. When the board found no issues, the neighbor tried to get the town selectmen to create a new bylaw banning poultry. When the townspeople voted the bylaw down, she says the selectmen pressured the Board of Health, which passed the bylaw as a "regulation," specifically targeting poultry and rabbits.

Kara hasn't given up the fight to keep her beloved ducks. FTCLDF is providing representation for her court challenge to the Board of Health's specious regulation.

Kara's situation is unfortunately common.

---

[13] Kara Zaks, "Town Health Board Limits Poultry But Not Cows," June 17, 2016, www.farmtoconsumer.org/blog/2016/06/17/town-health-board-limits-poultry-not-cows.

Taking Food Off a Family's Table

A similar thing happened to FTCLDF member Amber Bradshaw who is fighting for her family's livelihood over an ordinance prohibiting her from having chickens on their property.[14]

Amber has an amazing rags to riches story. She went from poverty to running a non-profit that provides healthy food and fresh produce to folks in need. The Bradshaw's home farm, run on a quarter of an acre in South Carolina, includes chickens and ducks as well as milking goats, bees, and a well-rounded garden.

Amber and her family complied with their homeowners association rule and other local rules. They never received a complaint. They were affected by an entirely separate dispute.

Two neighbors in another city got into a fight over backyard poultry. The city intervened and said that keeping poultry was "against the law."

There was no such law.

The poultry-owner and her lawyer then worked with the county council to pass a law "allowing" backyard chickens in five new zones out of the twenty-six zones in Amber's county.

Amber Bradshaw's land isn't in one of those zones.

She is a modern-day Robin Hood. At any time, the government could step in and take away their ability to put food on their table.

Amber says, "Every person should have the right to provide for their family, as long as we are not infringing on our neighbors' rights to do the same. This is wrong, and I will continue to fight."

FTCLDF is supporting her ongoing legal struggle.

Now might be a good time for a commercial message: FTCLDF relies on membership dues and other generous donations. Please support our efforts!

---

[14] Amber Bradshaw, "It's More Than Just Chickens—It's About Feeding My Family," May 27, 2016, www.farmtoconsumer.org/blog/2016/05/27/just-chickens-feeding-family.

First They Came for the Horses

But because there were no horses, they came after her chickens.

That's the story of FTCLDF member Stewart Goodwin who's been told by county officials in Virginia that she can't farm on her 100-year-old farming property after one anonymous neighbor took offense.

Stewart recently told her story at Joel Salatin's Polyface Farms during our annual Farm to Consumer Legal Defense Fund weekend. Her saga is also featured on our website.[15]

As Stewart tells it, "Because one unknown neighbor was offended by a sign I put in my yard, Henrico County has strongly suggested, under the threat of legal action, for me to remove my chickens from land that is zoned agricultural!"

The county first said her zoning is residential. It isn't.

They then told her she didn't meet setbacks for horses. She doesn't own any horses.

Then, they said she didn't meet setbacks to dwellings. A picture from the 1920s and a modern-day satellite image easily disproved that assertion. (Are we noticing a pattern?)

The county then changed its tune, going after her for lot lines and alleging that her poultry operation of thirty-six chickens is a threat to the general health, safety and welfare of the community. Can you believe it?

Thankfully, Virginia state law gives farmers like Stewart a right to farm.

Meanwhile, Virginia is also a Dillon's Rule state. Under Dillon's Rule, handed down by the Supreme Court of the United States, if there is a question about a local government's power or authority, then the local government does not receive the benefit of the doubt. Under Dillon's Rule, one must assume the local

---

[15] Stewart Goodwin, "No Chickens Allowed on the Farm," May 24, 2016, www.farmtoconsumer.org/blog/2016/05/24/no-chickens-allowed-farm.

government does not have the power in question.[16]

Even though Stewart's property began as a poultry operation in 1925, and even though it is zoned agricultural and is used for agriculture, the county continues to ignore state law in its ongoing proceedings against Stewart.

Stewart hasn't taken the matter sitting down. She got in touch with FTCLDF and got herself educated. Speaking as good as any lawyer, Stewart says that the application of land-use regulations to her property may be an improper "taking" (or inverse condemnation) by the government because the county is denying her reasonable and economically viable use of the land.

Right on!

Again, we have a situation where a landowner's right to farm has been violated.

FTCLDF will continue to support Stewart, as she so colorfully says, in her "continu[ing] fight to raise my chickens, to provide fresh, pesticide-free products to my customers and to protect the right of farmers and consumers to know where their foods originate without the fear of being poisoned by e. coli, listeria, salmonella or cancer-causing chemicals injected into our foods to preserve and extend the shelf life."

Right to Farm Laws in General

The unfortunate reality is that, even though FTCLF was initially formed to protect the rights of farmers and consumers to engage in direct commerce, and that remains a major focus of the organization's work, in recent years, we've spent an increasing

---

[16] Most states follow Dillon's Rule which says that municipalities only have power explicitly granted by the state. In contrast, Home Rule states like Maine, by statute or constitution, municipalities have almost unlimited power to legislate and regulate. In Home Rule states, municipalities may extend protection to farmers beyond that granted by an RTFA. "Dillon's Rule and Right-to-Farm," Vermont Natural Resources Council, vnrc.org/resources/community-planning-toolbox/land-use-law/right-to-farm/dillons-rule-and-right-to-farm (last visited Oct. 25, 2016). See also, infogalactic.com/info/Home_rule_in_the_United_States.

amount of time and resources fighting for the right of people to produce their own food.

Restrictive zoning ordinances have taken away a right that should be fundamental. Even on land that is zoned agricultural, local governments can take action to prohibit common agricultural activities if there are setback and other requirements for the activity.

There is also the problem documented above, of a neighbor using their influence to impose zoning and other restrictions on small farms. Then there is the dilemma of the overzealous bureaucrat reacting to an "anonymous" complaint.

One key tool in our fight against these types of legal encroachments is Right to Farm laws.

Right to Farm laws are on the books in all fifty states. However, the laws of each state vary somewhat. In general, they protect farmers from nuisance lawsuits and, in some cases, from zoning or other local ordinances.

In other states like Missouri, the state constitution itself has been amended to protect the rights of farmers. This after certain commercial farms took some substantial legal hits.

North Dakota was the first state to add such an amendment to its constitution. A constitutional amendment was rejected by voters in Oklahoma in 2016. The legislature of Maine, in 2015, as well as Nebraska and West Virginia, in 2016, considered constitutional amendments. The proposed amendment in West Virginia passed both houses but didn't make it to the ballot based on technical difficulties.[17] The Nebraska amendment may be reintroduced at a later date.[18]

With or without a constitutional amendment, state Right to

[17] E-mail from Dwayne O'Dell, West Virginia Farm Bureau, 15 November 2016, on file with author.
[18] Joanne Young, "Right-to-farm Resolution Pulled from Debate," Journal Star, Mar. 24, 2016, journalstar.com/legislature/right-to-farm-resolution-pulled-from-debate/article_40c9dff6-c62a-53d9-8fe9-4acd6a8bb005.html.

Farm laws were created to protect farmers specifically from *nuisance* lawsuits.

However, that doesn't mean one cannot be sued at all. Right to Farm may give farmers an affirmative defense or create a legal presumption that the farmer should prevail or be entitled to have the nuisance lawsuit quickly dismissed.

In certain states, however, even if the farmer has a defense under the Right to Farm Act, she still might have to prove it in court. For example, she might have to prove that her operations followed generally accepted agricultural practices, were nonnegligent, or wouldn't have been considered a nuisance before residents moved into the area.[19] Proving all that up can be time consuming and costly. Defending even against a frivolous lawsuit can be devastating to small farm operations.

Nevertheless, Right to Farm laws make lawsuits less likely to succeed. And a few states allow farmers to recover attorney fees and costs.

Right to Farm laws can also be used proactively. For example, when Louisiana banned burning due to dry conditions, farmers obtained an exemption because burning is a generally accepted agricultural practice protected by Right to Farm.[20] Meanwhile, in New York, a goat farm was able to turn a vacant barn in the center of town into a microcreamery using the Right to Farm statute.[21]

In addition to protecting a farmer from nuisance lawsuits, the laws in some states also protect farmers from changes to local

[19] *See, e.g., Trosclair v. Matrana's Produce, Inc.,* 717 So. 2d 1257, 1259 (La. Ct. App. 1998)(holding protection under RTFA based on GAAMP had to be proved by the farmer at trial as a question of fact)(produce operation sued by neighbor).
[20] Baylen Linnekin, "Right-to-Farm Debate Heats Up Controversies over laws in all 50 states that protect the rights of farmers to actually farm," Oct. 24, 2015, reason.com/archives/2015/10/24/right-to-farm-debate-heats-up. Linnekin is the co-author of the book, "Biting the Hands that Feed Us: How Fewer, Smarter Laws Would Make Our Food System More Sustainable" (2016) and is an expert on food law and safety: www.baylenlinnekin.com.
[21] *Id.*

ordinances!

As Rusty W. Rumley, a senior staff attorney for the National Agricultural Law Center, explains:[22]

> If your neighbor sells to a development and now there are 100 fami-lies living cross the road and they start complaining about odors coming from the farm, the Right to Farm statutes would protect you if they sued you directly for nuisance. But in states like Arkansas the law also will void local ordinances by saying the neighbors can't go to the city or county government and say, 'We are 100 families or 200 voters, and we want a new ordinance or will only elect a city council member who is going to take action against this farmer.'

Right-to-Farm: Model Act

The National Agricultural Law Center at the University of Ar-kansas has compiled Right to Farm laws state by state:

nationalaglawcenter.org/state-compilations/right-to-farm

Attorney Rumley, who works for that august organization, also published a helpful general comparison of the provisions found in Right to Farm statutes.[23]

Practicing attorneys should also know that there is an Amer-ican Law Reports Annotation that consolidates applicable cases from multiple jurisdictions.[24]

For our analysis of Right to Farm law, FTCLDF will take a slightly different approach.

First, we will analyze the American Legislative Executive Council's (ALEC) proposed model Right to Farm Act.[25] Many states based their legislation on this language. Even states that

[22] Allison Floyd, "Does Your State Have Right to Farm Protection?" May 12, 2014, growinggeorgia.com/features/2014/05/does-your-state-have-right-farm-protection.
[23] Rusty Rumley, A Comparison of the General Provisions Found in Right-to-Farm Statutes, 12 Vt. J of Environ. L 327(2013), available at vjel.vermont-law.edu/files/2013/06/A-Comparison-of-the-General-Provisions-Found-in-Right-to-Farm-Statutest.pdf.
[24] "Validity, Construction, and Application of Right-to-Farm Acts," 8 A.L.R.6th 465.
[25] American Legislative Executive Council, Model Right to Farm Act, Amended: Jan-uary 28, 2013, www.alec.org/model-policy/right-to-farm-act.

deviated from the model act are interesting in the manner of their deviation.

Let's (1) take a look at that model code. Then we will (2) discuss some key points that you might see in your own code. Finally, we will (3) include an appendix that summarizes the law state by state.

ALEC Model Act

The model act proposed by ALEC was intended to "provide for circumstances under which a farm shall not be found to be a public or private nuisance."

The model act states as follows:

A. A farm or farm operation shall not be found to be a public or private nuisance if the farm or farm operation alleged to be a nuisance conforms to generally accepted agriculture and management practices according to policy determined by the {Insert Appropriate State Agency}. Generally accepted agriculture and management practices shall be reviewed annually by the {Insert Appropriate State Agency} and revised as considered necessary.

B. A farm or farm operation shall not be found to be a public or private nuisance if the farm or farm operation existed before a change in the land use or occupancy of land within one mile of the boundaries of the farm or farm operation land, and if before that change in land use or occupancy of land, the farm or farm operation would not have been considered a nuisance.

C. A farm or farm operation is in accordance with subsection one of section two and shall not be found to be a public or private nuisance as a result of any of the following:
   1. A change in ownership or size.
   2. Temporary cessation or interruption of farming.
   3. Enrollment in government programs.
   4. Adoption of new technology.
   5. A change in the type of farm product being produced.

As with most civil law statutes, the model act has a separate section that defines its terms. For example, in defining what is or is not a "Generally Accepted Agricultural and Management Practices" (GAAMP) that would cause a farming operation to be protected, the model act defers to state agencies to make such determinations.

The model act also defines what constitutes a "farm" and provides a non-exclusive list of protected "farm operations," including but not limited to:

1. Marketing products at roadside stands or farm markets.
2. The generation of noise, odors, dust, fumes, and occasional conditions.
3. The operation of equipment and machinery necessary for a farm, including but not limited to irrigation and drainage systems and pumps and on-farm grain dryers, and the movement of vehicles, machinery, equipment, and farm products and associated inputs necessary for farm operations on the roadway as authorized by applicable motor vehicle laws.
4. Field preparation and ground and aerial seeding and spraying.
5. The application of chemical fertilizers or organic materials, conditioners, liming materials, or pesticides.
6. Use of alternative pest management techniques.
7. The fencing, feeding, watering, sheltering, transportation, treatment, use, handling, and care of farm animals.
8. The management, storage, transport, application and utilization of farm by-products, including manure or agricultural wastes.
9. The conversion from a farm operation activity to other farm operation activities.
10. The employment and use of labor.

Right to Farm Act: Key Provisions

Although Right to Farm Acts (RTFA) are on the books in every state and are enshrined in a couple state constitutions, and despite there being a model act available from ALEC, there are significant differences in the RTFA of each state.

In addition, while RTFAs cover agricultural activities it is less clear whether they cover nontraditional agricultural operations like agritourism including corn mazes, hay rides and the like.[26] Some state statutes explicitly cover agritourism while others may be open to interpretation by a court in the event of a lawsuit.

Nor does an RTFA serve as a defense to federal laws like the Clean Water Act.

Thus, the usual caveat applies: Your Mileage May Vary.

Moreover, official interpretation of a state's RTFA (by a judge or by an attorney general) can considerably affect the application of the act.

For instance, courts tend not to apply the RTFA where there has been a significant change in the farming operation, for instance, where traditional livestock raising techniques were converted to a concentrated animal feeding operation.[27]

At least one RTFA has been rendered practically useless by court interpretation. The RTFA in Iowa, which broadly extends nuisance immunity to all "agricultural areas,"[28] took a hit after the state supreme court declared certain parts of the RTFA unconstitutional. The Iowa supreme court held that farmers may have to compensate neighbors for nuisances.[29]

The Iowa Supreme Court held that immunity from bringing a nuisance action under the RTFA created an easement in the property affected by the nuisance. (An easement is a right to

---

[26] L. Paul Goeringer & H. L. Goodwin, University of Arkansas Division of Agriculture's Public Policy Center, "Arkansas' Right-to-Farm Law: An Overview," Paper No. FSPPC118-PD-2-13N, available at www.uaex.edu/publications/pdf/FSPPC118.pdf.

[27] See Payne v. Skaar, 900 P.2d 1352, 1355 (Idaho 1995).

[28] Iowa Code § 352.11 (LexisNexis 2016). Iowa is a "home rule" state wherein "Under chapter 352 the county board of supervisors is authorized to adopt a proposal for the creation or expansion of an agricultural area as submitted." Attorney General of the State Of Iowa, Opinion, 94-5-9, May 24, 1994, 1994 Iowa AG LEXIS 16.

[29] Bormann v. Board of Sup'rs In and For Kossuth County, 584 N.W.2d 309 (Iowa 1998) (county board of supervisor's decision to designate certain land as an "agricultural area" thereby bringing it under immunity to nuisance lawsuit under state's RTFA constituted a unconstitutional taking absent just compensation).

This decision isn't without controversy. See Eric Pearson, "Article: Immunities as Easements as "Takings": Bormann v. Board of Supervisors," 48 Drake L. Rev. 53 (1999).

cross or use someone's land in a certain way.)

According to that court, disallowing the nuisance suit under the RTFA deprived the neighbor of his ability to seek a remedy for the easement (use of his land by the offending farm) and, therefore, constituted an unjust taking under the state and federal constitution.

Practically speaking, a farm owner in Iowa may need to pay the adjoining property owners just compensation for any nuisance easement.

In summary, while an order to cease and desist won't necessarily be issued against a farming operation in Iowa, the farm would have to pay for the nuisance easement on surrounding land.[30]

That being said, the supreme court in the state of Idaho (and many other lower courts in other jurisdictions[31]) completely disagreed with the Iowa court's reasoning, saying that if a state legislature wants to take away someone's ability to bring a nuisance case in state court, it can do so.[32]

As one scholar noted, the competing rulings in Iowa and Idaho created a situation where, "In Iowa even the prospect of a farm nuisance is a taking, while in Idaho even the worst smoke and odor are not."[33]

---

[30] In 2004, the Iowa Supreme Court again stated "the statutory immunity [against a nuisance lawsuit] cannot constitutionally deprive property owners of compensation for the decreased value of their property due to the statutory imposition of an easement for the operation of an animal feeding operation as a nuisance. Because the recovery in diminution-in-value damages fully compensates the burdened property owners for the unlawful taking of an easement, the restrictions of the Takings Clause end at that point." *Gacke v. Pork Xtra, L.L.C.,* 684 N.W.2d 168 (Iowa 2004) (concentrated, large-scale animal feeding or confinement facility at issue).

[31] *See, e.g., Overgaard v. Rock County Bd. of Comm'rs,* 2003 U.S. Dist. LEXIS 13001 (D. Minn. July 25, 2003); *Labrayere v. Bohr Farms, LLC,* 2015 Mo. LEXIS 29 (Apr. 14, 2015); *Barrera v. Hondo Creek Cattle Co.,* 132 S.W.3d 544 (Tex. App. 2004).

[32] *Moon v. N. Idaho Farmers Ass'n,* 140 Idaho 536, 545 (Idaho 2004) (statute precluded nuisance as to smoke from burning).

[33] Adam Van Buskirk, Student Note, "Right-to-farm Laws as 'Takings' in Light of Bormann v. Board of Supervisors and Moon v. North Idaho Farmers Association," 11 Alb. L. Envtl. Outlook 169, 196 (2006). The Idaho statute in question dealing with burning has since been repealed.

If this sort of incredible juxtaposition drives you bonkers, welcome to my world. Just be glad you chose farming instead of law, though I'm not sure which of us shovels more BS in a day.

Point being: while RTFAs may be on the books, it's not enough to read the statutes. You must know how those statutes have been interpreted. That's why lawyers get paid the big bucks.

For example, if a neighbor tries to bring a nuisance lawsuit against you, a lawyer might ask whether that action was prompted by a change or an expansion of your farming operation.

On cursory review, courts take a dim view of legislative efforts to limit the rights of neighboring property owners, particularly where there's been some kind of change or expansion in farm operations.

Remember, RTFAs primarily protect farmers from residents who are "moving to the nuisance." Hence, RTFAs may not protect farmers who, themselves, create some additional nuisance.

That being said, ALEC's model code specifically protects farms that change their type of agricultural production, change size, or change technology. Many states adopted similar language.

As you can imagine, not every RTFA is created equal. For instance, some only limit private nuisances (a lawsuit by a specific, injured party) and not public nuisances (a lawsuit claiming damages to the public at large).

Nevertheless, it is possible to identify certain RTFA provisions that have proven to be helpful to small farmers and homesteaders.

If your state's RTFA has these provisions, you're in luck. If not, you might consider lobbying your state representative to see that your RTFA includes the following:

A. immunity from a nuisance suit if the farmer follows generally accepted agricultural and management practices (GAAMP) as defined by a state agency;

B. attorney fees and court costs for farmers defending against a
    nuisance suit;
C. provisions nullifying municipal ordinances to the contrary;
D. protection despite changes to the agricultural operation.[34]

Some RTFAs have other helpful provisions like robust legislative purposes sections; limitations on the types of damages
available; requirements for educating neighbors and notifying
new buyers as to right to farm.

Let's take a look at a few of these provisions in more detail
before turning to a state by state survey of Right to Farm laws.

(A) Generally Accepted Agricultural & Management Practices

Certain state RTFAs conclude that a nuisance does not exist, at
all, if generally accepted agricultural and management practices
(GAAMPs) are followed.

Other states give farmers a legal presumption if GAAMPs are
being followed, which would shift the burden of proof to the
plaintiff in a nuisance action. That's important because, as you
may know, it costs money to prove-up evidence in court. And if
a plaintiff can't offer sufficient proof, a defendant may seek to
have a case dismissed without ever having to submit any proof
of his own.

The next question, then, is who determines what's a "generally accepted agricultural and management practice?"

Some states don't say. In such cases, you'd have to fight it out
in court.

In Michigan, however, a committee decides which procedures are protected under the statute, while in New York such

---

[34] This list is adapted from Dowell's article: Tiffany Dowell, Student Note, "Daddy
Won't Sell the Farm: Drafting Right To Farm Statutes To Protect Small Family Producers," 18 S.J. Agric. L. Rev. 127, 133 (2009).

decisions are made on a case-by-case basis by the state department of agriculture[35].

In Michigan and New York, farmers can request that the state conduct an evaluation as to whether GAAMPs are being followed, thus triggering protection under the RTFA.

Getting "permission" ahead of time is better than risking a costly court trial. In addition, evaluations by the state agencies may be made available to other farmers who may be able to rely on those evaluations in their own farming operations.

If your state RTFA protects farmers who follow GAAMPs, it is important, then, to determine what is and what isn't a GAAMP in your locality and to find out how such a determination is made.

Michigan and New York allow for such determinations beforehand, which is great for risk-management. The RTFAs of other states are not so clear.

Wyoming, for instance, protects farmers who follow GAAMPs but doesn't define the term at all, which means a farmer will just have to argue his case in court.

Other states don't even include GAAMP language in their statutes, which can be even worse.

For instance, consider the legal horror faced by a cranberry farmer in Wisconsin who was sued by plaintiffs, including the state, alleging that the farm's use of phosphorous fertilizer damaged a nearby body of water from.

Despite being able to offer proof that his farm was following best operating practices as per the state university's guidelines, the farmer still faced a long, uphill, and prohibitively expensive

---

[35] N.Y. AGRIC & MKTS. LAW § 308 (LexisNexis 2016). A New York Court held that a case-by-case evaluative approach to determining which generally accepted agricultural practices are protected under the RTFA did not constitute a breach of a complainant's due process. Rather, the court deferred to the agricultural commission's authority, expertise and experience. *Pure Air & Water, Inc. v. Davidsen*, 246 A.D.2d 786, 668 N.Y.S.2d 248 (N.Y. App. Div. 1998) (hog farm's manure management program requested and obtained advisory opinion thereby preempting suit under RTFA).

legal battle.

The Wisconsin RTFA didn't contain a "generally accepted practices" provision. Instead, it required that an agricultural practice must have been in use before the change of local land use and that the agricultural practice not present a "substantial threat to public health or safety."

You are probably already shaking your head in anticipation of what happened next.

You guessed it: the case devolved into an incredibly expensive, semantic debate over whether his use of the fertilizer constituted a "substantial threat"[36] to public health.

We'll come back to the cranberry farmer's plight in the next few paragraphs. Suffice it to say, when reading your RTFA, it is crucial to determine whether there is a GAAMP provision and how and who decides what is or is not a generally accepted agricultural practice.

Finally, it must be noted that GAAMPs are a double-edged sword. On the one hand, an RTFA that includes GAAMPs can provide peace of mind and risk management provided you are able to comply with those practices.

However, in Michigan, the committee that defines GAAMP used its rule-making power to significantly limit the Right to Farm Statute.[37] A 2014 change to the Livestock Site Selection GAAMP removed Right to Farm protection from operations in primarily residential areas,[38] putting urban farmers and even those who want to raise vegetables for home consumption, at the mercy of local zoning boards!

For instance, an operation within 250 feet of a non-farm

---

[36] *State v. Zawistowski*, 309 Wis. 2d 233 (Wis. Ct. App. 2008).
[37] Rosemary Parker, "Michigan agriculture official clarifies new Right to Farm requirements," Michigan Live, May 06, 2014 at 11:50 AM, updated May 06, 2014 at 2:17 PM, available at www.mlive.com/news/kalamazoo/index.ssf/2014/05/michigan_agriculture_official.html.
[38] www.snopes.com/michigan-right-to-farm-act-repealed

house won't get GAAMP and therefore won't be protected under the RTFA. You can find more details under the Michigan section of our state-by-state Right to Farm survey.

This change to the GAAMPs in Michigan by an unelected and notoriously secretive committee made up of industry professionals was seen by some small farm advocates as a deliberate attempt to undermine the state RTFA and to reduce competition from small farms.[39]

A kinder reading of the change is that the RTFA was intended to prevent residential "moving to the nuisance" complaints and not to protect the opposite—that is, agricultural use "moving" into residential areas.

(B) Attorney Fees

Nineteen states allow recovery for attorney fees and costs to some degree.[40]

At the lowest end of the spectrum, Hawaii, Louisiana, Maine, Missouri, New Mexico, and Oklahoma say that a defendant "may" recover for "frivolous" lawsuits. South Dakota ups the ante by stating a defendant "shall" recover for "frivolous" suits. Meanwhile, North Carolina allows both a farming defendant to recover for a frivolous or malicious lawsuit and the plaintiff to recover if the farmer asserts a frivolous or malicious defense.

It is extremely difficult to prove that a lawsuit was "frivolous" or "maliciously" brought.

At an intermediate level, Arkansas and Michigan state that a prevailing defendant "may" recover attorney's fees and costs based on the court's discretion. Washington is similar but also allows for recovery of exemplary damages where a lawsuit was

---

[39] Randy Zeilinger, "What we Heard: Meeting of the Michigan Commission of Agriculture, January 27, 2016," 1/30/2016, www.michigansmallfarmcouncil.org/blog/what-we-heard-meeting-of-the-michigan-commission-of-agriculture-january-27-2016.
[40] Please see the state survey below for additional information and specific citations.

brought maliciously and without cause. Meanwhile, Kansas allows a prevailing defendant to recover for suits based on improper use of agricultural chemicals.

Finally, certain states have strong wording as concerning fees and costs. For example, Illinois, Iowa, and Wisconsin say that prevailing defendants "shall" receive attorney's fees and costs. Oregon adds fees and costs on appeal as well. In Illinois, it's further been noted that "prevailing" means receiving an actual judgment in your favor, not if you settle or end the case through some other mutual agreement. Hence, any settlement agreements in those states should take attorney's fees and costs into consideration.

Meanwhile, in Delaware, a prevailing defendant in an Agricultural Preservation District "shall" be awarded attorneys' fees and costs.

Similarly, in New York, attorney fees and costs shall be awarded to a farming defendant if the agricultural practice was sound according to an opinion issued by the commissioner prior to the start of any trial or settlement of such action.

The Texas Right to Farm Act contains a broad statement that a person who sues an agricultural operation "is liable" to the agricultural operation for attorney's fees and costs.

Provision for attorney's fees and costs is very important.

For example, in the cranberry farm case mentioned above, the farmer was able to take his case all the way to the state supreme court, where he prevailed, [41] because his insurance company understood he could recover attorney fees and costs if he won. Otherwise, that farmer may never have been able to afford to defend himself.[42]

---

[41] He actually succeeded at the appellate court level and the supreme court denied the petition to review that decision. *State v. Zawistowski*, 309 Wis. 2d 233 (Wis. Ct. App. 2008).

[42] Tiffany Dowell, Student Note, "Daddy Won't Sell the Farm: Drafting Right To Farm Statutes To Protect Small Family Producers," 18 S.J. Agric. L. Rev. 127, 142-43 (2009).

For many small farmers, the threat of a lawsuit by itself can be enough to put them out of business.

Fortunately, members of FTCLDF receive legal support from our organization. We encourage you to become a member.

If your state RTFA does not include an attorney fee provision, we also recommend that you contact your state legislators to see about adding a provision to provide for the recovery of attorney fees and court costs.

(C) Negating Municipal Ordinances

In the experience of FTCLDF, the nullification of local ordinances is a key protection provided by Right to Farm statutes. Yet, not every RTFA explicitly contains such provisions. Moreover, some provisions are more strongly worded than others.

The RTFAs in 25 states, to some degree, make conflicting ordinances null and void. Illinois and Tennessee courts have done the same by court decision. This brings the number to 27 states that have some protection against intrusive zoning ordinances and other regulations. In addition, it has been suggested that the constitutional amendment in Missouri will also have the effect of voiding contrary ordinances.[43]

Taking a look at specific provisions, the law in Michigan is quite restrictive. In 1999, Michigan amended its RTFA[44] to exempt farming operations from local zoning ordinances that

---

[43] Please see the state survey below for more information and for specific citations.
[44] Mich. Comp. Laws § 286.474(6) (LexisNexis 2016)

> [E]xcept as otherwise provided in this section, it is the express legislative intent that this act preempt any local ordinance, regulation, or resolution that purports to extend or revise in any manner the provisions of this act or generally accepted agricultural and management practices developed under this act. Except as otherwise provided in this section, a local unit of government shall not enact, maintain, or enforce an ordinance, regulation, or resolution that conflicts in any manner with this act or generally accepted agricultural and management practices developed under this act.

would extend or revise "in any manner" the act or generally accepted agricultural and management practices (GAAMPs).

Based on this amendment, a court in Michigan overturned a local zoning ordinance that prohibited the raising of farm animals on property of less than three acres. The court said the ordinance conflicted with the state RTFA because the ordinance tried to prevent a protected farm operation by limiting the size of the farm.[45]

In subsequent cases, the courts in Michigan have protected farmers from similar zoning incursions.

However, a recent revision to state GAAMPs removed Right to Farm protection from many residential areas, thereby giving zoning boards free rein.[46]

Some scholars support that decision. They say that preventing municipalities from controlling public health and welfare within their localities goes too far, particularly as CAFOs replace traditional farms.[47]

For instance, a few years back, a township in Pennsylvania was found to lack authority to adopt ordinances regulating the use of sewage sludge to fertilize a tree farm. The court said applying sewage sludge using a no-till method that substantially increased the smell, was nevertheless a "normal agricultural operation" that was regulated under other state law provisions. Hence, it was protected under the Right to Farm act.[48] In that case, other farming neighbors brought the nuisance action. See the Pennsylvania section of the state survey for a grisly description of the case.

On the other hand, in Alaska, the RTFA simply says, "The

---

[45] *Charter Twp. of Shelby v. Papesh*, 267 Mich. App. 92 (2005)(holding that poultry were "farm products" and raising poultry constituted a "farm operation" that could not be precluded by township ordinance).

[46] www.snopes.com/michigan-right-to-farm-act-repealed.

[47] Wendy K. Walker, "Whole Hog: The Pre-Emption of Local Control by the 1999 Amendment to the Michigan Right to Farm Act," 36 Val. U.L. Rev. 461, 462 (2002).

[48] *Commonwealth v. East Brunswick Twp.*, 980 A.2d 720 (Pa. Commw. Ct. 2009).

provisions of [the RFTA] supersede a municipal ordinance, res-olution, or regulation to the contrary."[49] The Alaska Supreme Court interpreted this provision to apply only to "nuisance ac-tions" and not to unrelated local ordinances, like moving a fence so the municipality can repair an adjacent road.[50]

Similarly, Connecticut's RTFA preempts private or public nuisance actions "notwithstanding any general statute or munic-ipal ordinance or regulation pertaining to nuisances to the contrary." The courts in Connecticut, however, held that zoning ordinances like setback requirements or requirements for filing and getting approval of a land management plan are not "nui-sance ordinances" and, thus, are not void under the RTFA.[51]

Conflicts between an RTFA and zoning ordinances often de-volve into technical legal questions as the following two examples show.

All the Kings Horses

Does every animal in America belong to the State? Like Robin Hood, must we face the Sherriff of Nottingham for hunting the king's deer? It sure seems like it sometimes.

For instance, a farm in the town of Redding, Connecticut found itself on the wrong end of an injunction. The local zoning board said the farmers couldn't have more than nine horses without getting approval for a land management plan.[52] The town only allowed one horse per lot without a plan.

However, the farm had those horses prior to ordinance in question. Hence, their nine horses were a perfectly legal, "prior

---

[49] Alaska Stat. § 09.45.235 (LexisNexis 2016).
[50] *Gates v. City of Tenakee Springs*, 822 P.2d 455 (Alaska 1991). The landowner had been granted a permit to build the fence in front of her property on the public right of way. Due to a storm, the city ordered all encroachments removed so they could repair the adjacent road.
[51] *See e.g., Ammirata v. Zoning Bd. of Appeals*, 65 Conn. App. 606 (2001), affirmed on re-mand *Ammirata v. Zoning Bd. of Appeals*, 81 Conn. App. 193 (2004).
[52] *Id.*

nonconforming use" that was grandfathered in.

The farmers fought the injunction and received a stipulated judgment requiring them to submit an expensive land management plan only if they maintained more than nine horses on their property.

Problem solved, right?

Wrong.

As happens in too many cases, the zoning commission took another tack.

While the farm was fighting the initial injunction, the municipality issued a cease and desist order. The town claimed the farm needed to comply with a twenty-five foot paddock setback. And, again, the town required plaintiffs to file a land management plan.

The farmers cried foul, saying that they didn't have to file the land management plan unless they maintained more than nine horses. The farmers also complained that the zoning commission should've raised the setback requirement when it brought the other injunction.

The courts disagreed. Unlike a civil litigant or a criminal prosecutor, a zoning board does not have to raise every issue all at once so long as "there is no finding or record of piecemeal enforcement resulting in harassment of a property owner."

The courts don't seem to consider that a zoning commission has the unlimited resources of the state while a farm's resources are quite limited in comparison.

If you feel like you're getting piecemealed or harassed, document those occurrences and make sure your attorney enters that evidence into the record.

At trial, the court recognized that the RTFA negated "any general statute or municipal ordinance or regulation pertaining to nuisances to the contrary."

However, the court held that the setback ordinance wasn't a "nuisance ordinance." Thus, the horse farm had to comply with the twenty-five foot setback requirement since they hadn't offered evidence that their paddock predated that requirement.

Let me get technical for a paragraph or two. It is extremely concerning that the RTFA might not even void an ordinance like the following Redding town ordinance. Redding has an ordinance that prohibits "Any use which results in contamination of air, ground, water or the natural environment, beyond the specific limits prescribed below" as well as "Any use which is noxious by reason of emission odor, dust, gases, smoke, noise, vibration, light, radiation, or danger of explosion or other physical hazard."

That looks suspiciously like the RTFA's list of activities that *will not* constitute an actionable nuisance.

The Connecticut court, however, held that Redding's ordinance might *not* be characterized as a "nuisance ordinance," prohibited under the RTFA, but as a proper exercise of a town's police power to manage public health and safety.[53]

Similarly, going back to the case at hand, the court held requiring approval of a land management plan is not a "nuisance ordinance" but is part of the town's proper exercise of police power.

A land management plan requires extensive and costly filings including: "a map of the site and its immediate environs, showing the extent of the proposed operations, approximate slopes and drainage patterns, general location of streams, wetlands, buildings, [fence lines] and roadways, and shall be accompanied by a written description which explains in detail how the operations shall be conducted, parties responsible, and demonstrates that sound land management practices will be adhered to consistent with [the law]."

---

[53] See the state survey on Wisconsin for discussion on a similar case: *Town of Trempealeau v. Klein*, 870 N.W.2d 247 (Wis. Ct. App. 2015) (unpublished decision), available at cases.justia.com/wisconsin/court-of-appeals/2015-2014ap002719.pdf.

After the land management plan is filed, the zoning commission may or may not give the plan a public hearing. And the commission may (or may not) approve the plan, or approve it with stipulations.

As to the farm in Redding with nine horses, the court said that the requirement of a land management plan doesn't interfere with their use of the land. The town agreed it could not prohibit the continuation of a valid nonconforming use (having nine horses on the land) and the court surmised that the requirement to file a land management plan would not cause the farmer's nonconforming use to be lost.

One wonders, however, whether their rights will, as a practical matter, be lost due to the stifling cost of preparing and submitting such a plan? Moreover, what happens if or when the commission decides not to approve the plan?

The court in Connecticut narrowly interpreted the RTFA's provision that limits zoning ordinances.

Thankfully, other states have broader language, as we learned when a county board of supervisors tried to put an oyster farm in Virginia out of business.

Pearls Before Swine

In Virginia, the Right to Farm Act proved crucial in a four-year legal battle to protect a farmer's right to operate an oyster business on his land.[54]

FTCLDF member Anthony Bavuso's legal troubles began when the county board of supervisors said he had to have a special use permit to conduct his oyster farm. Because he had his farm and residence on the same property, the county zoning administrator issued a determination letter saying that Bavuso

---

[54] Farm to Consumer Legal Defense Fund, "FTCLDF Member Bavuso Wins Virginia Right to Farm Case," July 18, 2015, www.farmtoconsumer.org/blog/2015/07/18/ftcldf-member-bavuso-wins-virginia-right-to-farm-case.

needed a special use permit.

Bavuso applied for the special use permit while appealing the determination letter.

Surprise, surprise. The board of supervisors refused to issue the special use permit. And the board of zoning appeals denied his appeal.

Bavuso argued all the way to the Virginia Supreme Court that the county zoning code didn't require that he have a special use permit. After the Supreme Court rejected that argument, Bavuso argued successfully that the zoning code violated the Right to Farm Act. In July 2015, the circuit court agreed, ruling in Bavuso's favor. That case is now on appeal to the state supreme court.

This case illustrates the importance of having the RTFA protect farmers from local ordinances, particularly ordinances sought by a rich or influential neighbor who objects to agricultural practices.

If your state's RTFA contains such a provision, it may be crucial in defending your Right to Farm.

But, as we saw with the horse farm case in Connecticut, the wording of such a provision is also critical.

However, even if your state has a weakly worded provision or none at all, that may not be the end of the story. For instance, the Illinois RTFA does not contain language nullifying local zoning ordinances. However, an appellate court nullified an ordinance anyway. Under Illinois law, municipalities may not adopt ordinances that infringe upon the spirit of the state law or are repugnant to the general policy of the state. The ordinance banned all commercial farming in a village. The court said the ban violated state policy to preserve agricultural land and also violated the spirit of the RTFA to protect farms from becoming nuisances based on a change in surrounding use.[55]

FTCLDF or a local lawyer may be able to help you get legal relief from a pernicious local ordinance based on the RTFA, particularly if there have been technical or procedural violations by the zoning board or local officials.

If you are an FTFLDF member, document everything. Please give us a call as soon as you get wind of a problem like this. If you're not a member, please consider joining up for a small, yearly membership fee!

(D) Changes or Expansions

The model Right to Farm act promoted by the American Legislative Executive Council (ALEC) suggests that farms should not be considered a nuisance based on a change in ownership or size; or changes based on a cessation or interruption in farming or enrollment in a government program; nor changes based on new technology; or changes based on the type of farm product being produced.

Nevertheless, as many as 16 states have no explicit language protecting expansions or changes.[56] Among those states, a court in Massachusetts found that some changes may nevertheless be protected.[57] However, the courts of many states, including Nebraska have held that the RTFA only protects farms from

---

[55] *Vill. of Lafayette v. Brown*, 27 N.E.3d 687 (Ill. App. 2015), appeal denied *Vill. of Lafayette v. Brown*, 392 Ill. Dec. 370 (Ill. 2015). But see *Town of Trempealeau v. Klein*, 870 N.W.2d 247 (Wis. Ct. App. 2015) (unpublished decision), available at cases.justia.com/wisconsin/court-of-appeals/2015-2014ap002719.pdf, where the court was not impressed with legislative-purpose type language.

[56] These states include Alabama, Arizona, California, Hawaii, Illinois, Maryland, Massachusetts, Montana, Nebraska, New Jersey, New York, Rhode Island, North Dakota, Utah, and Wyoming. Tennessee is also included; however, I believe there to be a strong argument from statutory construction that expansions and new technology are protected.

[57] *Francisco Cranberries LLC vs. Edward P. Gibney, Jr.*, 1999 Mass. App. Div. 223 (Mass. App. Div. So. 1999).

changes in and around the locality of the farm (i.e. coming to the nuisance), and not changes to the farm itself.[58] These case decisions highlight the importance of protecting reasonable changes and expansions under the RTFA.

Several states do just that. For instance, at least six states have language similar to the ALEC model act including Arkansas, Colorado, India, Kansan, Michigan, and North Carolina.

Michigan, for example, seems to have adopted the model act nearly verbatim.[59] Likewise, Indiana's RTFA is substantially similar, protecting farming operations absent a "significant change."[60] In Indiana, a significant change does not include changes in the type of agricultural operation, changes in size or ownership, change in status in a government program, or the use of new technology.[61]

A handful of states, like Alaska, provide protection regardless of any subsequent expansion or adoption of new technology.[62] Other states that allow changes or expansions include Idaho, Iowa, Oklahoma, West Virginia, South Carolina, and South Dakota. In some states, like Minnesota, expansions or even substantial alterations are allowed but they may reset the clock on RTFA protection (many states require the operation to have existed for one, two, or even three years before protection vests). In states like Washington, Washington only changes in the type of product and the use of new technology may be protected.

Although changes and expansions may appear to be covered under the RTFA, a farmer must still be cautious, as case law in

---

[58] *Flansburgh v. Coffey*, 370 N.W.2d 127 (Neb. 1985); *Cline v. Franklin Pork, Inc.*, 361 N.W.2d 566 (Neb. 1985).
[59] Mich. Comp. L. Servs. § 286.473 (LexisNexis 2016).
[60] Ind. Code § 32-30-6-9 (LexisNexis 2016).
[61] *Id.*
[62] Alaska Stat. § 09.45.235 (2016).

Idaho illustrates. Idaho courts found that the RTFA did not protect changes to a cattle feedlot[63] and, later, to a hog feedlot[64] because the "changes [came] from the offending nuisance, not from changes in the character of the surroundings."[65] However, in 2011, the legislature amended the RTFA to include protection for "expansions" to agricultural facilities or operations. Nevertheless, the Idaho Supreme Court reaffirmed its earlier rulings finding that a horse-riding arena built next to a neighbor's house constituted a nuisance.[66] The court found that the "change [came] from the offending nuisance, not from changes in the character of the surroundings." (Remember, RTFAs were created to protect farms from residential neighbors *moving to* the nuisance.)

Some states attempt to draw a balance when it comes to changes in farming operations.

In Georgia, for instance, a farming operation that has been in operation for one year or more is protected under the RTFA. Moreover, that one-year timing requirement does not reset based on an expansion of physical facilities or adoption of new technology.[67]

Hence, if an existing farming operation expanded in Georgia, and a neighbor sued, claiming that the expansion constituted a nuisance, the farm would seem to have a defense under the RTFA.

Not so fast, says the courts in Georgia. "[T]hat which may constitute a nuisance regardless of urban sprawl, such as polluting a stream [or presumably a substantial change in farm operations], is never protected by the statute since such activity

---

[63] *Payne v. Skaar*, 127 Idaho 341 (1995) (cattle feedlot tripled in size over twenty years).
[64] *Crea v. Crea*, 16 P.3d 922 (Idaho 2000).
[65] *McVicars v. Christensen*, 156 Idaho 58, 63 (2014).
[66] *McVicars v. Christensen*, 156 Idaho 58, 63 (2014).
[67] Ga. Code Ann. § 41-1-7 (LexisNexis 2016).

does not become a nuisance as a result of 'changed conditions in the surrounding locality.'"[68]

Again, the Georgia court decision demonstrates that RTFAs were created to protect farmers from encroaching residential use. RTFAs may not protect farmers if the change in the farming operation itself is the source of the nuisance.

Minnesota's RTFA does the exact opposite, resetting the time clock with each significant alteration or subsequent expansion "by at least 25 percent in the number of a particular kind of animal or livestock located on an agricultural operation." [69] Nevertheless, "significant alterations" do not include things like: a change in family ownership; temporary cessation or interruption of cropping activities; adoption of new technologies; or a change in the type of crop produced.[70]

Pennsylvania also resets the clock for "substantial expansions or alterations." However, that would not include things like "new activities, practices, equipment and procedures" so long as those are "consistent with technological development within the agricultural industry."[71]

Moving on, the protection afforded by some states is quite broad, including Wisconsin, which protects farmers regardless of whether "a change in agricultural use or agricultural practice is alleged to have contributed to the nuisance."[72] Mississippi and Oregon are similar in this regard.

Finally, there's a subset of states whose RTFAs warn that while some changes may be allowed, "significant" or "substantial" expansions or changes may not be protected. These states include Connecticut, Delaware, Florida ("more excessive farm operation" not protected), Kentucky, Louisiana, Maine ("material change" after a change in local use negates protection), New

---

[68] *Herrin v. Opatut*, 248 Ga. 140, 142 (1981)(comment in brackets added by author).
[69] Minn. Stat. § 561.19(2016).
[70] *Id.*
[71] 3 Pa. Stat. § 952 (LexisNexis 2016).
[72] Wis. Stat. Ann. § 823.08(LexisNexis 2016).

Hampshire, New Mexico, Texas, and Vermont.

Missouri takes a detailed approach. It allows for the usual changes in ownership as well diminutions in size or temporary cessations. But the Missouri RTFA also allows farming operations to reasonably expand provide they comply with local, state, and federal environmental codes, laws, and regulations and provided the expansion doesn't have a substantially adverse effect on the environment, or create a hazard to public health or "a measurably significant difference in environmental pressures upon existing and surrounding neighbors because of increased pollution."[73]

As you can see, RTFAs vary state by state as to the types of expansions or changes that remain protected under the act.

The elephant in the room, of course, is the conversion of traditional farming practices into CAFOs. Neighbors, even neighboring farmers, may not appreciate the mundungus scents that emanate from commercial swine, cattle, or poultry operations.

Several states now break out CAFOs separate and apart from their RTFA.

Regardless, farmers who are contemplating a significant change to their farming operation should seek legal advice to make sure the change doesn't remove protection under the RTFA.

Although FTCLDF may be able to give general advice to our members, sometimes the best advice is to secure a competent local attorney who can advise you as to the specific laws of your state as well as court and local administrative decisions.

FTCLDF can help members locate a good, local lawyer who will be sympathetic to the needs of small farms and homesteaders.

Having summarized Right to Farm, and having analyzed the model code as well as specific, important provisions, we now

---

[73] Mo. Rev. Stat. § 537.295 (2016).

turn to a state by state survey of the Right to Farm law.

Right to Farm Law: State Survey

Although this section is intended to be comprehensive as of Winter 2016, due to the nature and sheer volume of legislation and regulations, ours is an aspirational goal.

It's not possible to summarize everything.

For instance, almost every RTFA includes a definition section that defines what is or isn't a farming operation.

Some of these definitions include things like agrotourism, aquaculture, silviculture, and even farmers' markets and roadside stands. Others do not.

We've tried to point out where a definition section includes something unique or unexpected. We haven't summarized every definition section.

Moreover, we've tried to highlight where a particular state's RTFA addresses zoning ordinances; but we can't cover every possible agricultural zoning exemption here. Some exemptions, like zoning exemptions for agricultural barns and structures are beyond the scope of this resource.

Also, it's possible we missed something. Please let us know if you're aware of something critical that has been omitted: info@farmtoconsumer.org.

As always: YMMV and, when in doubt, contact FTCLDF directly or seek the advice of a good lawyer in your locality.

Alabama: Alabama Code § 6-5-127[74]

An agricultural establishment or farming operation facility should have been in operation for more than one year without a court finding it to be a nuisance. If so, no private or public nuisance can attach based on "changed conditions in and about the locality thereof," unless there is negligent or improper operation. The statute doesn't preclude suits for damages relating to water pollution or overflow.

In addition, as part of the Family Farm Preservation Act,[75] the Alabama legislature voids local ordinances and resolutions that would treat farms as a public or private nuisance.[76] There are some limitations, especially for CAFOs. However, the average FTCLDF member should be protected from local ordinances provided you've been operating for a year and your operation is being lawfully run in accordance with generally accepted agricultural and farm management practices.[77] It's not clear who defines such practices.

In a published case dealing with the RTFA, the Alabama Supreme Court held that where a neighbor was residing on adjacent property before a commercial chicken farm operation began and where there had been no "change in conditions in and about the locality," the RTFA didn't apply.[78] In that case it was the farm that changed, not the neighboring use. It wasn't as though nonagricultural use was encroaching on agricultural lands; quite the opposite—a change in the use of agricultural land was affecting residential neighbors.

It may interest you to know that the Alabama Farmers Association has a book available online titled "Ag Law for You" that

---

[74] codes.findlaw.com/al/title-6-civil-practice/al-code-sect-6-5-127.html.
[75] Ala. Code § 2-6B-1 to -7 (LexisNexis 2016)
[76] Ala. Code § 2-6B-5 (LexisNexis 2016).
[77] Ala. Code § 2-6B-3 (LexisNexis 2016).
[78] *Swedenberg v. Phillips*, 562 So. 2d 170 (Ala. 1990).

covers many areas including right to farm.[79]

Alaska: Alaska Statutes § 09.45.235[80]

The legislature in Alaska says that an agricultural facility is not and does not become a private nuisance as a result of a changed condition in the area if the agricultural facility was not a nuisance at the time the agricultural facility began agricultural operations.

The time agricultural facility began agricultural operations refers to the date on which any type of agricultural operation began on that site regardless of any subsequent expansion of the agricultural facility or adoption of new technology.

The RTFA doesn't apply to flooding or to improper, illegal, or negligent conduct of agricultural operations. However, it does supersede any municipal ordinance to the contrary.

That being said, the Supreme Court of Alaska held that the RTFA defends against *nuisance* ordinances, not against a permit revocation under city ordinances, for example, requiring a landowner to move a fence that had previously been permitted.[81]

You would need to carefully examine whether a conflicting ordinance was a "nuisance" ordinance or an independent exercise of the locality's police power.[82]

---

[79] alfafarmers.org/uploads/files/Ag_Law_Book_-_Final_-_January_2013.pdf.
[80] codes.findlaw.com/ak/title-9-code-of-civil-procedure/ak-st-sect-09-45-235.html.
[81] *Gates v. City of Tenakee Springs*, 822 P.2d 455 (Alaska 1991). The landowner had been granted a permit to build the fence in front of her property on the public right of way. Due to a storm, the city ordered all encroachments removed so they could repair the adjacent road.
[82] *See, e.g., Town of Trempealeau v. Klein*, 870 N.W.2d 247 (Wis. Ct. App. 2015) (unpublished decision), available at cases.justia.com/wisconsin/court-of-appeals/2015-2014ap002719.pdf; *Ammirata v. Zoning Bd. of Appeals*, 65 Conn. App. 606 (2001), affirmed on remand *Ammirata v. Zoning Bd. of Appeals*, 81 Conn. App. 193 (2004).

Arizona: Arizona Code § 3-112[83]

Arizona is home to the *Spur Industries*[84] case, arguably the most famous farm nuisance case. It came about when, in the early 1970s, a residential subdivision sprung up alongside a commercial cattle feedlot. The court stopped the feedlot's operations but required the residential real estate developer to indemnify the farmer for his losses. This is one of the original "coming to the nuisance" cases and highlights the need for Right to Farm laws.

Arizona's RTFA, adopted in 1990, reads:

A. Agricultural operations conducted on farmland that are consistent with good agricultural practices and established prior to surrounding nonagricultural uses are *presumed to be reasonable* and do not constitute a nuisance unless the agricultural operation has a substantial adverse effect on the public health and safety.

B. Agricultural operations undertaken in conformity with federal, state and local laws and regulations *are presumed* to be good agricultural practice and not adversely affecting the public health and safety.

For example, in 2006, where a farmer followed state regulations relating to agricultural composting, an appellate court held that the legal presumption afforded under the RTFA applied.[85]

If your farming practice isn't covered by federal, state, or local laws, you may have to prove in court that it is "consistent with good agricultural practices." (You wouldn't get a cost-saving, motion-to-dismiss-prompting legal presumption.)

Arkansas: Arkansas Code § 2-4-107[86]

Arkansas seems to have a fairly strong RTFA. It protects farmers

---

[83] www.azleg.state.az.us/FormatDocument.asp?inDoc=/ars/3/00112.htm&Title=3&DocType=ARS.

[84] *Spur Industries v. Del E. Webb Dev. Co.*, 108 Ariz. 178 (1972). The farmer was permanently enjoined from operation as a public and private nuisance; but the plaintiff, having brought people to the nuisance, was ordered to indemnify defendant for damages caused by the injunction.

[85] *Coconino County v. Antco*, 214 Ariz. 82, 85 (Ariz. Ct. App. 2006).

[86] law.justia.com/codes/arkansas/2010/title-2/subtitle-1/chapter-4/2-4-107.

from public or private nuisance actions provided (1) they've been in business for a year or more prior to the change in local land use that gives rise to the complaint and provided (2) they "employ methods or practices that are commonly or reasonably associated with agricultural production."

Actually, the statute gives farmers a rebuttable presumption provided they employ "methods or practices that are commonly or reasonably associated with agricultural production or are in compliance with any state or federally issued permit."

As to changes or expansions, the Arkansas RTFA follows the American Legislative Executive Council's model code protecting the following: changes in ownership or size; nonpermanent cessations or interruptions of farming; participation in any government-sponsored agricultural program; the employment of new technology; or changes in the type of agricultural product produced.

Finally, the RTFA says that the court "may" award attorney's fees to the prevailing party, which could include the person bringing the suit if they were to prevail.

As to local ordinances, Arkansas Code § 2-4-105 states that any local or county ordinances that attempt to make agricultural operations a nuisance are void. Although there are no cases interpreting this provision, the state attorney general's office has issued positive opinions stating in one instance that "a city would have no jurisdiction to adopt ordinances regulating livestock auction barns for control of odor or noise, but the city would still retain the power to regulate according to the public health statutes."[87] The state attorney general has also noted that the one-year timing requirement protects agricultural activities that have been in operation for a year from any change in local rules.[88]

---

[87] Ark. Op. Att'y Gen. No. 83–194 (1983).
[88] *See* Ark. Op. Att'y Gen. No. 86-199 (1986).

California: California Civil Code § 3482.5[89]

It shouldn't surprise anyone that the situation in California is particularly complex.

In addition to the RTFA, which was added to the civil code in 1981,[90] as of the turn of the century more than 40 counties and 50 cities had adopted their own right-to-farm ordinances.[91] Some local ordinances purport to give a farmer even *more* protection than the RTFA.[92] Moreover, cities and counties are allowed to require realtors to give notice to potential buyers moving into agricultural areas.

The state-wide RFTA says a farm should meet seven steps in order to qualify for immunity from nuisance. The activity in question must be (1) agricultural activity (2) conducted or maintained for *commercial* purposes (3) in a manner consistent with proper and accepted customs and standards (4) as established and followed by similar agricultural operations in the same locality. Moreover, the nuisance claim must arise (5) due to any changed condition in or about locality (6) after the agricultural activity has been in operation for more than *three years*; and the activity (7) must not have been nuisance at time it began.

Agricultural activity protected under the act includes but isn't limited to: the cultivation and tillage of the soil; dairying; the production, cultivation, growing, and harvesting of any agricultural commodity including timber; viticulture, apiculture, or horticulture; the raising of livestock, fur bearing animals, fish, or poultry; and any practices performed by a farmer or on a farm as incident to or in conjunction with those farming operations,

[89] leginfo.legislature.ca.gov/faces/codes_displaySection.xhtml?lawCode=CIV&sectionNum=3482.5.
[90] California Civil Code § 3482.5 (LexisNexis 2016).
[91] Matthew Wacker, Alvin D. Sokolow & Rachel Elkins, "County Right-to-Farm Ordinances in California: An Assessment of Impact and Effectiveness," University of California Agricultural Issues Center, AIC Issues Brief, No 15 (May 2001), available at aic.ucdavis.edu/oa/brief15.pdf.
[92] *Id.*

including preparation for market, delivery to storage or to market, or delivery to carriers for transportation to market.

Unique to California is the 3 year timing requirement. The commercial purpose requirement is a bit unique; several other states have similar requirements.

As to the 3 year timing requirement, some local municipalities have dropped that to 1 year by ordinance. And at least one California court has upheld this additional protection.[93]

Moreover, a change in ownership won't reset the 3 year clock. The applicability of the statute is *predicated on duration of the agricultural operations*, not the duration of farmland's ownership.[94]

The nuisance exception provided by the RTFA isn't unlimited. An agricultural activity can't "obstruct the free passage or use, in the customary manner, of any navigable lake, river, bay, stream, canal, or basin, or any public park, square, street, or highway."

In addition, the nuisance exception does not invalidate "any provision contained in the Health and Safety Code, Fish and Game Code, Food and Agricultural Code, or Division 7 of the Water Code if the agricultural activity would constitute a nuisance under one of those codes."

You should be fine as long as you're operating "in a manner consistent with proper and accepted customs and standards as established and followed by similar agricultural operations in the same locality" even if that means setting off deafening propane cannons to deter birds from eating the grapes off your vines.[95]

---

[93] *Ad Hoc Committee for Clean Water v. Sonoma County Bd. of Sup'rs*, 2002 WL 1454105 (Cal. App. 1st Dist. 2002).

[94] *Rancho Viejo LLC v. Tres Amigos Viejos LLC*, 123 Cal. Rptr. 2d 479 (Cal. App. 4 Dist. 2002).

[95] "CPD states it cannot take action against propane cannons," Record-Bee, posted 08/08/14, 12:01 AM PDT, www.record-bee.com/article/ZZ/20140808/NEWS/140808019.

On another positive note, California's RTFA *does* overrule local ordinances. Specifically, the RTFA "shall prevail over any contrary provision of any ordinance or regulation of any city, county, city and county, or other political subdivision of the state." As aforementioned, that doesn't mean a local ordinance can't provide *even more* protection.

The RTFA doesn't preclude a city or county from "adopting an ordinance that allows notification to a prospective homeowner that the dwelling is in close proximity to an agricultural activity."

As to changes or expansions, if the nuisance is based on changes to your farm property, you might not be protected under the RTFA. For example, a farmer changed from rice to row crops, then complained that overflow from remaining, neighboring rice farms caused a nuisance to his row crop farm. The statute precludes nuisance actions "due to any changed condition in or about the locality." In that case, a court said it was the plaintiff's change to row crops that constituted the "changed condition" creating a nuisance. The defendants' use, rice farming, hadn't changed.[96]

Looking at it from the perspective of the rice farmers. They had been in operation for three years, then, based on a changed condition in the locality (the neighbor's change to row cropping) they were being sued for nuisance. This is the exact situation the RTFA protects against.

For agricultural *processing* activities, the state of California affords immunity from nuisance actions similar to the RTFA. However, as to processing activities that "substantially increase," nuisance actions can be brought "with the respects to those increases in activities or operations that have a significant effect on the environment."[97]

However, activities that have been in effect for more than

---

[96] *Souza v. Lauppe*, 69 Cal. Rptr. 2d 494 (Cal. App. 1997).
[97] Cal. Civ. Code & 3482.6 (LexisNexis 2016).

three years get a rebuttable presumption that the increase was not substantial.

The RFTA itself, on the other hand, is silent as to "substantial increases." This is problematic for farmers. It's very significant where a legislature could have said something, and didn't. Here the statute on *processing* activities mentions "substantial increases." The RTFA doesn't. The negative corollary, then, is that substantial increases in agricultural *activity* may not be protected under the RTFA.

Moving on, in California, a neighbor can't try to recharacterize a nuisance lawsuit as a "trespass" claim in order to avoid the RTFA.[98]

As to cases in California, a court upheld a city's decision declaring that an emu farm was a public nuisance. The bird farm consisted of 800 ostriches and emus on a ranch located in a residential community within city limits. A year later, the city declared the operation to be a public nuisance. The couple challenged the city's decision. But a court held against the couple saying, first, they hadn't offered any evidence that their operation was "conducted in a manner consistent with accepted standards, as established by similar agricultural operations in the same locality." Second, although they'd had birds on the ranch intermittently for 12 years, nothing in the record indicated that the present 800 bird operation had existed for the statutory minimum of 3 years.[99] It is interesting to note that the farmers had the burden of proving all the elements in order to establish immunity under the RTFA.

In another case, a commercial farmer subdivided his land, sold portions of the property to a residential developer, but continued to farm on the rest. A court held that the residential developer couldn't sue the farmer in nuisance after irrigation

---

[98] *Rancho Viejo LLC v. Tres Amigos Viejos LLC*, 123 Cal. Rptr. 2d 479 (Cal. App. 2002) (holding property damage allegedly sustained from irrigation water qualified as "nuisance" within literal language of statute).
[99] *Mohilef v. Janovici*, 58 Cal. Rptr. 2d 721 (Cal App. 2nd Dist. 1996), review denied.

water damaged the developer's property. The developer wanted to argue that the farmer's conduct wasn't reasonable. The court said the reasonableness of the farmer's conduct wasn't material to the application of the statute.[100] The pertinent question under the statute is whether the commercial agricultural activity at issue (the commercial farmer's irrigation) was an accepted and customary practice followed by similar operations in the locale.[101] The court found that it was and held for the farmer.

Finally, in a recent case, a downslope neighbor sued a lemon grove for damage after a rainstorm flooded the downslope property with water, mud, and debris. A jury verdict of $350,000 was set aside by the trial court based on the RTFA. However, on appeal, the higher court said a new trial should be held because it wasn't factually clear from the record whether the lower court had found the lemon grove to have been operated in a manner consistent with proper and accepted customs and standards.[102] If it wasn't, then then RTFA protection wouldn't apply.

Colorado: Colorado Revised Statutes Annotated § 35-3.5-102[103]
Agricultural operations are not public or private nuisances where they were in operation prior to the non-agricultural use that gave rise to the alleged nuisance. Agricultural operations must also employ "methods or practices that are commonly or reasonably associated with agricultural production" and not be negligently operated.

Agricultural operations that follow common methods or those reasonably associated with agricultural production get a

---

[100] A lengthy discussion of this point can be found in *Gilbert v. Synagro Cent., LLC*, 131 A.3d 1, 20 (Pa. 2015) and in the Pennsylvania portion of the state survey below. As I understand it, in some states, operating in a negligent manner would be grounds for voiding RTFA protection. The only question in California, however, is whether you've met the seven elements, none of which mention negligence.

[101] *Rancho Viejo v. Tres Amigos Viejos*, 100 Cal App 4th 550 (2002).

[102] *W&W El Camino Real, LLC v. Fowler*, 226 Cal. App. 4th 263 (2014).

[103] codes.findlaw.com/co/title-35-agriculture/co-rev-st-sect-35-3-5-102.html.

rebuttable presumption that they are *not* operating negligently.

Common or reasonable agricultural methods or practices aren't defined by statute or by administrative agency. So, a farmer who is sued in negligence may have to prove that his methods and practices are "commonly or reasonably associated with agricultural production."

As for changes or expansions, the RTFA includes language similar to the ALEC model code, allowing for changes in ownership, non-permanent cessation or interruption of farming, participation in any government sponsored agricultural program, employment of new technology, or a change in the type of agricultural product produced.

Attorney fees and costs "may" be awarded by the court to the prevailing party.

Local governments may provide even *more* protection than the RTFA (unless they try to prevent an owner from selling or hinder the owner in seeking approval to put the land into alternative use).

However, local ordinances and resolutions cannot make an agricultural operation a nuisance.

For instance, it was unreasonable and violative of the RTFA for a county to institute an absolute prohibition on the movement of agricultural sprinklers on a county road.[104]

Connecticut: Connecticut General Statutes § 19a-341[105]

Provided they've been in operation for one year or more, haven't substantially changed, and follow generally accepted agricultural practices (GAAMP), Connecticut farms are immune from private or public nuisance actions "notwithstanding any general

---

[104] *Bd. of County Comm'rs of Logan County v. Vandemoer*, 205 P.3d 423 (Colo. App. 2008).
[105] www.ct.gov/doag/cwp/view.asp?a=1366&q=259086.

statute or municipal ordinance or regulation pertaining to nuisances to the contrary."

As of 2011, as many as 12 Connecticut towns had adopted
right-to-farm ordinances, though these mostly parrot the state
RTFA.[106]

As for determining what is or isn't GAAMP, the statute states
that "inspection and approval of the agricultural or farming operation, place, establishment or facility by the Commissioner of
Agriculture or his designee shall be prima facie evidence that
such operation follows generally accepted agricultural practices."

Moreover, the statute provides a specific list as to what isn't
a nuisance including odor, noise, dust, chemicals used in an approved fashion, or water pollution done in an approved fashion
except where it pollutes public or private drinking water supplies.

As you will see from the cases below, it is important to obtain
inspection and approval so as to create a prima facie case that
your operations comply with GAAMP.

For example, one farm had to go to trial where a neighbor
said the farm did not follow generally accepted farming practice.
The neighbor claimed the farmer housed more cows on their
farm than permitted under the special exception, stored too
much manure for too long a time under the waste management
plan, and failed to replace or repair broken pipes that were necessary and important to the proper functioning of the waste
disposal plan. The judge held that the case could not be automat-
ically dismissed, because whether the farm violated GAAMP
was a question of fact for court to decide at trial.[107]

In another case, a court overturned a local ordinance. The city

[106] Kristen L. Miller, "Right to Farm Ordinances," OLR Research Report, 2011-R-0058,
Jan. 31, 2011, available at www.cga.ct.gov/2011/rpt/2011-R-0058.htm (naming
Brooklyn, Canterbury, Colchester, Columbia, Eastford, Franklin, Lebanon, New Mil-
ford, North Stonington, Pomfret, Thompson, and Woodstock).
[107] *Pestey v. Cushman*, 1994 Conn. Super. LEXIS 3275, 6-7 (Conn. Super. Ct. Dec. 15,
1994).

gave legal notice that a farmer had to move the fence line of his horse corral 200 feet away from a nearby residence. However, the court refused to enforce the city's order!

The court noted that the RTFA protects farming activities that comply with GAAMP. The department of agriculture had issued a report stating that the farm complied with GAAMP. Therefore, the local ordinances couldn't be applied to the farm.

On the other hand, the court ordered the farmer to abandon an unused well inside his horse corral because the RTFA doesn't protect the pollution of public or private water supplies.[108]

As you can see, it is very important in Connecticut to get GAAMP protection. According to the a toolkit published by the Connecticut Farm Bureau Association, GAAMPs can be found via the Connecticut Department of Agriculture, the Connecticut Department of Energy and Environmental Protection, the United States Department of Agriculture (USDA) Natural Resources Conservation Service(NRCS), and the University of Connecticut (UCONN) Cooperative Extension System.[109]

That guide goes on to remind Connecticut farmers that the RTFA doesn't protect them in the event of negligence or willful or reckless misconduct. Neither does it exempt them from complying with local planning and zoning regulations in general. Nor does it preempt municipal zoning regulations in general. The courts have said that the RTFA only applies to urban sprawl and only specifically negates "nuisance ordinances."

In contrast to the horse corral case described above, where GAAMP protection prevented application of a local ordinance, a farm in Redding, Connecticut was ordered to comply with a twenty-five foot paddock setback and to file a land management plan. The court held that these requirements were not "nuisance

---

[108] *Havlicek v. Hills*, 2003 Conn. Super. LEXIS 3298 (Conn. Super. Ct. Dec. 5, 2003).
[109] Connecticut Farm Bureau Association, "A Tool Kit for Connecticut Farmers: Right to Farm Law" (2014) available at www.cfba.org/images/resources/right_to_farm.pdf.

ordinances" but were proper exercises of municipality's police power.[110]

Similarly, where a farm used helicopters in conjunction with agricultural activity, a court nevertheless upheld an injunction based on a local ordinance. Presumably the ordinance was meant to prevent noise. However, nuisances arising from farm noise are explicitly precluded by the RTFA, which also voids nuisance ordinances. Regardless, the court upheld a local ordinance that enjoined helicopters from coming and going from or being stored on farm property. The court said the purpose of the Act was to protect agricultural operations from urban and residential encroachment whereas the ordinance in question merely prohibited the defendants from launching and landing helicopters on their land.[111] Presumably the court felt the ordinance was not a "nuisance" ordinance but was a proper exercise of the city's police power.

Had the farm secured inspection and established GAAMP concerning the use of helicopters, one wonders if the case would have gone differently.

Oh, apparently the RTFA won't protect you if you try to get back at a neighbor by maliciously sticking an animal pen, a box truck, five sheep, and two goats directly behind their property line.[112] Go figure! (You might be surprised at how many similar cases we came across.)

Delaware: 3 Delaware Code § 1401; 10 Delaware Code § 8141; 3 Delaware Code § 910

In Delaware, if you've been in compliance with all applicable

[110] *Ammirata v. Zoning Bd. of Appeals*, 65 Conn. App. 606 (2001), affirmed on remand *Ammirata v. Zoning Bd. of Appeals*, 81 Conn. App. 193 (2004).
[111] *Town of Enfield v. Enfield Shade Tobacco, LLC*, 32 Conn. L. Rptr. 240 (Conn. Super. Ct. 2002).
[112] *Morytko v. Westfort*, 2005 Conn. Super. LEXIS 1456 (Conn. Super. Ct. May 31, 2005).

state and federal laws, regulations, and permits and offer proof that you've been in operation for one year or more, you get an absolute defense to a nuisance action[113]—provided you conform to federal, state or local health or zoning requirements and don't operate in a negligent or improper manner.[114]

Furthermore, compliance with all applicable state and federal laws, regulations, and permits (1) creates a *legal presumption* that the operation was conducted in a manner consistent with good agricultural practice, and (2) prevents a state or local law enforcement agency from bringing a criminal or civil charge against you.[115]

As for local ordinances, "Any and all ordinances of any unit of local government now in effect or hereafter adopted that would make the operation of any such agricultural operation or its appurtenances a nuisance or providing for abatement thereof as a nuisance in the circumstances set forth in this section and *shall be null and void.*"[116]

That being said, you may lose RTFA protection in Delaware (1) if you do not conduct your operation in a manner consistent with "good agricultural practice" or (2) when there has been a significant change in the operation itself.

Good agricultural practice is legally presumed when one is "in compliance with all applicable state and federal laws, regulations, and permits."

Finally, there is a special exemption for agricultural operations located in Agricultural Preservation Districts.[117] The code preempts nuisance suits against normal agricultural uses and activities conducted in accordance with good husbandry and best management practices in Agricultural Preservation Districts, "including any such claims under any existing or future county

---

[113] 3 Del. Code § 1401 (LexisNexis 2016).
[114] 10 Del. Code § 8141 (LexisNexis 2016)(defining agricultural operation).
[115] 3 Del. Code § 1401 (LexisNexis 2016).
[116] 10 Del. Code § 8141 (LexisNexis 2016).
[117] 3 Del. Code § 910 (LexisNexis 2016).

or municipal code or ordinance."

Moreover, an owner sued for nuisance in an Agricultural Preservation District, if she prevails "shall" be awarded attorneys' fees and costs.

We could find no cases applying the RTFA in Delaware.

Florida: Florida Statute § 823.14[118]

A farm operation that has been ongoing for a year or more is protected from public or private nuisance suits, provided the operation conforms to generally accepted agricultural and management practices (GAAMP) and wasn't a nuisance at the time operations began.

The statute lists several exceptions to RTFA protection including: untreated or improperly treated waste, dead animals, or harmful gasses as well as improper septic tanks and the like; keeping diseased animals outside a federal disease control program; and unsanitary places where animals are slaughtered.

As to changes or expansions, neither a change in ownership, a change in the type of product, a change in conditions around the locality of the farm, or a change that is made in order to comply with local, state or federal best management practices will deprive one of protection under Florida's Right to Farm Act.

That being said, the act specifically doesn't protect "a more excessive farm operation with regard to noise, odor, dust, or fumes" where the farm was next to an established business or homestead on March 15, 1982.

Expansions are a bit more tricky. Expansions to land boundaries will create a new "established date of operation" for each expansion. This means it would take one year before farm operations on that land would be protected. However, that does not remove protection for the preexisting operation. And expanded

---

[118] www.leg.state.fl.us/Statutes/index.cfm?App_mode=Display_Statute& URL=0800-0899/0823/Sections/0823.14.html.

operations *within* the same boundary retain the original established date of operation.

Florida has strict rules nullifying local ordinances or regulations. Because it intends to eliminate duplication of regulatory authority over farm operations, except as otherwise provided, no rule, policy, or regulation can be adopted "to prohibit, restrict, regulate, or otherwise limit an activity of a bona fide farm operation *on land classified as agricultural land*" pursuant to the authority of the department environmental protection, or the department of agriculture and consumer services, or water management districts.

Note that you have to be on land classified as agricultural land and be a bona fide farm operation. Moreover, this rule only affects ordinances and regulations created after the passage of the RTFA.[119]

In addition, the Florida code exempts non-residential farm buildings from the Florida building code and any county or municipal building code. A "non-residential farm building" means any building or support structure that is used for agricultural purposes, is located on a farm that is not used as a residential dwelling, and is located on land that is an integral part of a farm operation or is classified as agricultural land.[120]

Moving to court cases interpreting the RTFA, where there was a substantial change to the way a farm handled waste disposal, a court allowed a nuisance case to proceed. In that case, a large chicken operation changed from a dry to a wet manure distribution process and increased frequency of collection. Applying the language of the statute, the court held there had been a "substantial change" on a farm with an adjacent residence or business as of March 1982 and, therefore, the county's attempt to enforce waste disposal ordinances against the chicken farm

---

[119] *Wilson v. Palm Beach County*, 62 So.3d 1247 (Fla. App. 2011).
[120] Florida Statute § 604.50 (LexisNexis 2016).

might be allowed despite the RTFA.[121]

The court said that the RTFA is not "an unfettered license for farmers to alter the environment of their locale merely because the practices which they used in 1982 were acceptable at that time."[122]

However, a minor odor change or minimal degradation of the local environment may not constitute a "more excessive farm operation" that would remove protection under the RTFA. Whether a substantial change took place is a fact-intensive question determined on a case-by-case basis.

In a different case, a court held that a county has to give a farmer "due process," which means giving him a chance to prove that his activities are protected by the RTFA. In that case, the county said a farmer was operating an unauthorized "junkyard" on his agricultural zoned land by "storing" a bushhog, a bulldozer, a crane, a backhoe, and various other equipment and materials outdoors at his farm. But the county didn't allow the farmer to call witnesses to prove that the machinery was used for agricultural purposes. Hence, a Florida court overruled the county's order.[123]

The attorney general's office in Florida has also interpreted the RTFA as it applies to local ordinances. In general, non-residential farm buildings aren't subject to local zoning ordinances.[124] However, the AG's office said that non-residential farm buildings *may* be subject to zoning regulations that don't "prohibit, restrict, regulate, or otherwise limit an activity of the farm." In that case, a setback requirement for building construction would not necessarily limit a farm's operation.[125]

---

[121] *Pasco County v. Tampa Farm Serv., Inc.*, 573 So. 2d 909 (Fla. App. 1990).
[122] *Id.* at 912.
[123] *Kupke v. Orange County*, 838 So. 2d 598 (Fla. Dist. Ct. App. 5th Dist. 2003).
[124] Florida Statute § 604.50 (LexisNexis 2016).
[125] Advisory Legal Opinion, AGO 2001-71, Oct. 10, 2001, available at www.myfloridalegal.com/ago.nsf/Opinions/4386A18F9B5715E085256AE1006783B7.

Local zoning ordinances were again upheld by the AG in a case where a farmer optimistically claimed a "barn" was a "non-residential farm building," even though the building had a bathroom, a kitchen, and bedrooms where the farmer's children and guests could stay.[126]

Finally, the AG said that a county could enforce regulations against a dairy farm. In that case, the dairy farm changed its fertilization process. The farm began fertilizing with waste using a center pivot irrigation system. Applying the language of the RTFA statute, the AG said if there was an adjacent residence or property as of March 1982 and if the fertilizing practice represented a changed, "more excessive" operation that involves significant or substantial degradation in the locale, the county may enforce regulations applicable to those changes.[127] The negative implication being that, otherwise, the neighbors were literally SOL.

Georgia: Georgia Code Annotated § 41-1-7[128]

No agricultural operation, including the operation of any roadside market, can be a public or private nuisance as a result of changed conditions in or around the locality, provided it has been in operation for one year or more and where the complained of nuisance isn't the result of negligent, improper, or illegal operation.

As to the one year timing requirements, neither expanded facilities nor changes in technology will reset the clock.

The RTFA is fairly straightforward in its application. For example, the state attorney general issued an opinion saying that

---

[126] Advisory Legal Opinion, AGO 2009-26, June 15, 2009, available at www.myfloridalegal.com/ago.nsf/Opinions/F4901B3D5EA1BD34852575D6006C857D.
[127] Advisory Legal Opinion, AGO 2006-07, Mar. 8, 2006, www.myfloridalegal.com/ago.nsf/Opinions/A6785134F928A9A68525712C00565C84.
[128] codes.findlaw.com/ga/title-41-nuisances/ga-code-sect-41-1-7.html.

an egg farm that is located in a nonagricultural, residential area is not protected from nuisance suits.[129]

Moreover, a state court refused to apply RTFA protection to a chicken farm that expanded from 40,000 to 500,000 chickens. The neighbors complained of odors and flies. Waste drainage from the farm allegedly killed fish in a neighboring pond. The court said that the RTFA did not apply because the nuisance was caused, not by urban sprawl intruding upon the farm, but by a change on the farm itself. RTFA protection is given where the nuisance is "a result of changed conditions in or around the locality."[130]

The implication is that a significant change to an existing farming operation may remove RTFA protection. Forewarned is forearmed.

Hawaii: Hawaii Revised Code § 165-4[131]

Hawaii's code states: "No court, official, public servant, or public employee shall declare any farming operation a nuisance for any reason if the farming operation has been conducted in a manner consistent with generally accepted agricultural and management practices. There shall be a rebuttable presumption that a farming operation does not constitute a nuisance."

It is not clear how "generally accepted agricultural and management practices" (GAAMP) are defined.

On a side note, farming operations in Hawaii include aquaculture.[132]

Recovery of attorneys' fees and costs are available for frivolous lawsuits.[133]

---

[129] 1980 Op. Att'y Gen. No. U80-51.
[130] *Herrin v. Opatut*, 248 Ga. 140 (1981).
[131] www.capitol.hawaii.gov/hrscurrent/Vol03_Ch0121-0200D/HRS0165/HRS_0165-0004.htm.
[132] Hawaii Rev. Code § 165-3 (LexisNexis 2016).
[133] Hawaii Rev. Code § 165-5 (LexisNexis 2016).

Idaho: Idaho Code § 22-4503[134]

An agricultural operation or facility (or an expansion thereof) does not become a private or public nuisance based on changes in surrounding nonagricultural use, provided it's been in operation for one year and provided the operation or expansion wasn't a nuisance at the time it began or was constructed. However, improper or negligent operation is not protected.

As to pertinent definitions, "Improper or negligent operation" means that the agricultural operation is not undertaken in conformity with federal, state and local laws and regulations or permits, and adversely affects the public health and safety. Meanwhile, protected agricultural operations include aquaculture and roadside markets.[135]

As to expansions, it must be remembered that the RTFA applies to nuisances caused by "a change in surrounding nonagricultural use" like encroaching urbanization. Changes or expansions on a farm itself that create a nuisance may not be protected.

Case in point, the Idaho Supreme Court held against expansions of a cattle feedlot[136] and, later, of a hog feedlot.[137] The court said the RTFA applies to the encroachment of "urbanizing areas" and to situations where there have been changes in "surrounding nonagricultural activities." The changes in those cases were with the feedlots themselves.

Although Idaho's RTFA was amended in 2011 to include expansions, the Idaho Supreme Court affirmed its earlier rulings, holding against a horse farm that erected a fabric, indoor horse riding arena next to a residence. In that case, the "changes [came] from the offending nuisance, not from changes in the character

---

[134] legislature.idaho.gov/idstat/Title22/T22CH45SECT22-4503.htm.
[135] legislature.idaho.gov/idstat/Title22/T22CH45SECT22-4502.htm.
[136] *Payne v. Skaar*, 127 Idaho 341 (1995) (cattle feedlot tripled in size over twenty years).
[137] *Crea v. Crea*, 16 P.3d 922 (Idaho 2000).

of the surroundings."[138]

Despite the dangers of encroaching residential use to agricultural operations, a court held that neighboring farms could not
prevent a fellow farmer from dividing his property into five lots,
four of which would be used for residential purposes. The neighboring farmers argued that the RTFA was created to discourage
"premature removal of land from agricultural uses." However,
the court said that the neighbors' agricultural operations would
not be affected by the new residential development since the
zoning board required the residential developer to include right
to farm marketing disclosures and right to farm deed restrictions.[139] (On the other hand, a court in another state upheld
a zoning board's refusal to plat a residential neighborhood next
to a large orchard.[140] The facts showed the residential neighbors
simply couldn't live next to the farming operation without pos-
sibly being damaged by noise and chemicals.)

In response to concerns about litigious social justice groups,
Idaho has a law on the books that criminalizes "interference with
agricultural production." That criminal law punishes people
who trespass or who pretend to be employees in order to "ex-
pose" an agricultural production facility by taking surreptitious
pictures or recordings or in order to do real, actual damage.[141]
However, that state law was held to be unconstitutional by a federal district court.[142] That decision has been strongly criticized
and may not hold up on appeal.[143]

---

[138] McVicars v. Christensen, 156 Idaho 58, 63 (2014).
[139] Whitted v. Canyon County Bd. of Comm'rs, 137 Idaho 118 (2002).
[140] Id.
[141] Idaho Code § 18-7042 (2016).
[142] Animal Legal Def. Fund v. Otter, 118 F. Supp. 3d 1195 (D. Idaho 2015).
[143] Western Watersheds Project v. Michael, 2016 U.S. Dist. LEXIS 88843 (D. Wyo. July 6,
2016) Criticizing the Idaho court's decision, the court in Wyoming said, "There is no
precedent to support the premise that whistleblowers somehow have an elevated
First Amendment right to make audiovisual recordings on private property without
permission. No matter how virtuous or important one may view a whistleblower's
motives or actions, the ends do not justify the means of trespass." Id. at *28.

Illinois: 740 Illinois Compiled Statutes 70/1 to /5[144]

The RTFA in Illinois protects farms from public or private nuisance lawsuits based on changed conditions in the surrounding area that occur more than one year after the farm has been in operation, provided the farm wasn't operating in a negligent or improper manner.

The term "farm" is quite broad, applying to "any parcel of land used for the growing and harvesting of crops; for the feeding, breeding and management of livestock; for dairying or for any other agricultural or horticultural use" in any combination.[145]

The nuisance exemption doesn't apply to pollution or changed conditions of the waters of any stream or to any overflow of lands.

A defendant who prevails in the courts *shall* receive attorney fees and costs. Not if you settle or end the case through some other mutual agreement. Only if you get an actual judgment in your favor.

Moreover, although the statute does not explicitly void contrary ordinances, an appellate court held that the RTFA prevents a "municipality through an ordinance [from depriving] the owner of the right to use the land to grow crops, either agricultural or horticultural."[146]

We will now turn to a detailed analysis of two key cases.

In interpreting the RTFA, the Illinois Supreme Court in a case called *Toftoy v. Rosenwinkel*[147] held that where an adjacent farmhouse residence is purchased next to an ongoing farm operation, the change in ownership can constitute a "changed condition"

[144] www.ilga.gov/legislation/ilcs/ilcs3.asp?ActID=2034&ChapterID=57.
[145] 740 Ill. Comp. Stat. Ann. 70/2 (LexisNexis 2016).
[146] *Vill. of Lafayette v. Brown*, 27 N.E.3d 687 (Ill. App. 2015), appeal denied *Vill. of Lafayette v. Brown*, 392 Ill. Dec. 370 (Ill. 2015).
[147] *Toftoy v. Rosenwinkel*, 983 N.E.2d 463 (2012), available at www.illinoiscourts.gov/opinions/SupremeCourt/2012/113569.pdf.

in the locale so as to prevent the residents from bringing a nuisance action.

The trial and appellate court had said that because the farmhouse had continually been a residence, there had been no change in the surrounding locale as required by the language of the act. Therefore, the RTFA would not apply.

The supreme court reversed, holding that the change in ownership represented a "changed condition" that was very much like "coming to the nuisance." The buyers moved into the neighboring house; they started complaining. The supreme court said the new farmhouse owners couldn't bring a nuisance action against the cow farm.

In 2015, an appellate court interpreted the *Toftoy* case. The appellate court decision is important not only for its holding on the facts, but also because the court held that zoning ordinances are negated by the RTFA.

That case involved farmers who bought a bankrupt nursery that had been growing trees and prairie grass. The new owners began growing corn and soybeans. Ten months later, the village passed an ordinance prohibiting all commercial farming.

There appeared to be two legal hurdles. First, the Illinois RTFA does not include language that negates zoning ordinances.

Second, even if the RTFA negates ordinances, did the farm itself qualify for protection under the precise language of the RTFA.

Turning to the first point, the appellate court refused to enforce a village ordinance that banned all commercial farming.[148]

The court stated that the Illinois RTFA "limits the ability of either an individual through private suit *or a municipality through an ordinance* to deprive the owner of the right to use the land to grow crops, either agricultural or horticultural."[149]

The court reasoned that the purpose of the RTFA is to protect

---

[148] *Vill. of Lafayette*, 27 N.E.3d at 694.
[149] *Id.* at 694.

and prevent the loss of agricultural land by *limiting the circumstances under which farming operations may be deemed to be a nuisance.*[150]

Looking to previous case law, the court said, "It is well established that municipalities may not adopt ordinances which infringe upon the spirit of the state law or are repugnant to the general policy of the state."

In Illinois, a local ordinance that infringes upon the legislative intent of a state statute is preempted.

In this case, the RTFA says that a farm shall not be deemed a nuisance if it satisfies the requirements of the Act. However, the ordinance labeled all commercial farming a nuisance. The ordinance contained no exception for farm operations that satisfied the elements of the Act. Therefore, the court held that the ordinance was preempted.[151]

Not so fast, wrote another judge. (Appellate courts have at least three judges, and each judge may choose to agree with the majority's written opinion, or write a concurring opinion, or disagree by writing what we call a "dissenting opinion.") In order to qualify for protection under the RTFA, the farming operation still had to meet all the elements of the RTFA.

For instance, the one year requirement.

The court decided that the one year requirement had been satisfied because a commercial farming operation had been on the land for more than a year. The change in ownership or changing the crop from trees and prairie grass to corn and soybeans did not reset the one year timing requirement. The court said that removing protection in such cases would negate state policy to preserve farmland.

The court asked rhetorically: who would buy a farm inside a

---

[150] Not every state has found such aspirational language in the statute to be binding so as to negate municipal ordinances. See the discussion under Wisconsin of the case *Town of Trempealeau v. Klein*, 870 N.W.2d 247 (Wis. Ct. App. 2015) (unpublished decision), available at cases.justia.com/wisconsin/court-of-appeals/2015-2014ap002719.pdf.

[151] *Id.* at 694-95.

municipality knowing that a zoning ordinance passed within a year could put them out of business?

One judge disagreed. Should a new buyer assume a right to change the farming operation? That judge concluded that a *significant change* in the farming operation, like changing from row farming to a commercial hog farm, might negate protection under the RTFA.[152] However, as that judge noted, the ordinance in question did not refer to the type of change but banned *all* commercial farming within the village and was, thus, illegitimate.[153]

In addition, the RTFA requires that the nuisance be based on "changed conditions in the surrounding area" such as residences encroaching on farm land.

Based on this plain language, the courts in other jurisdictions found that changes on the farm land itself that become nuisances to residential neighbors are not protected.[154] The RTFA protects farms only against "coming to the nuisance" lawsuits.[155]

---

[152] *Id.* at 695.

[153] *Id.* at 695-96.

[154] For example, interpreting a similarly-worded statute, the Idaho Supreme Court held that a fabric, indoor horse riding arena built adjacent to neighbor's property wasn't protected by the RTFA because the "changes [came] from the offending nuisance, not from changes in the character of the surroundings." *McVicars v. Christensen*, 156 Idaho 58, 63 (2014).

[155] The concurring opinion points to the following cases:

See e.g., *Flansburgh v. Coffey*, 220 Neb. 381, 370 N.W.2d 127, 130-31 (Neb. 1985) (holding that the Nebraska Right to Farm Act did not apply where the change at issue occurred on the defendant's farm rather than on other land in the vicinity of the farm); *Crea v. Crea*, 135 Idaho 246, 16 P.3d 922, 925 (Idaho 2000) (holding that defendant's expanded hog farm operations were not protected by the Idaho Right to Farm Act because the hog operation was alleged to be a nuisance "not because of changes in surrounding non-agricultural uses, but because of an expansion of the [farm] operation itself"); *Durham v. Britt*, 117 N.C. App. 250, 451 S.E.2d 1, 3-4 (N.C. Ct. App. 1994) (ruling that North Carolina's "right to farm" law did not apply to "situations in which a party fundamentally changes the nature of the agricultural activity which had theretofore been covered under the statute," and holding that a "fundamental change" occurred where the defendant, who previously operated turkey houses, changed his farming operation to a hog production facility (emphasis in original)).

The village, therefore, argued that there had been no "changed conditions in the surrounding area." The change that prompted the town's action was the changes on the farm. Nothing had changed with the village.

The court disagreed, citing to the previously discussed *Toftoy* case. In that case, as you may recall, a couple bought a neighboring farmhouse, then complained about the adjacent farm. The state supreme court held that the change in ownership (i.e., a change in legal rights) represented a "changed condition" for the purposes of the act.

The appellate court reasoned that, like *Toftoy*,[156] "a change in a local ordinance can likewise be a 'changed condition'" that triggers RTFA protection.

The ordinance represented a change in legal rights or legal status and was, therefore, a change in the surrounding area that would allow the RTFA to apply.

The court noted, something "provoked" the village to pass the ordinance months after the farm was purchased. The farm itself had always been adjacent to residential property and had always been used for commercial farming purposes. "The crop raised at the commercial farming operation simply changed," says the court. Thus, the change must have been to the sensibilities of the village itself which now sought to ban all commercial farming.

To quote the court: "Obviously, some nearby residents did not like the new crop and all that goes with planting and harvesting it; the Village enacted the ordinance."

Applying the rationale of *Toftoy*, defendants' commercial farming endeavors could not "become a nuisance" to the Village

---

Even more cases could have been listed, like *Powell v. Tosh*, 929 F. Supp. 2d 691 (W.D. Ky. 2013) (expansion of hog farming operations not protected by RTFA because no change to local lands, the neighbors had always lived there).
[156] *Toftoy v. Rosenwinkel*, 983 N.E.2d 463 (2012).

until the Village amended the ordinance. That change to the ordinance was a "changed condition" that gave rise to the Village's nuisance action.

It could be argued, *per contra*, that the change in farm operations created a nuisance to which the village was responding. To use the court's logic: the adjacent residential property *use* didn't change, there must have been a change to the farm that gave rise to the nuisance claim. As aforementioned, the courts of other jurisdictions have not applied the RTFA in such instances. Therefore, it is not clear how much precedential value this Illinois Appellate Court case will have.

Moving on to another point, the village also argued that because the statute lacked the ALEC model code language protecting farms in the event of a cessation or a change in the type of agriculture, those changes were not protected under the act. Their technical, legal argument is that because those words are included in other state RTFAs and in the ALEC model code and yet are not found in the Illinois statute, it should be assumed the Illinois legislature departed from the other statutes and from the model act on purpose.

The majority opinion disagreed, stating that the purpose of the RTFA "is to protect and conserve the development and improvement of agricultural land and limit, not expand the circumstances under which farms can be considered nuisances" (such as a change in the type of crop or temporary cessation especially due to foreclosure).

Such rhetorical language is a little hard to follow.

Perhaps the court's analysis could kindly be restated as follows: The enumeration by other states of significant changes that won't affect application of their RTFA does not mean lack of such language in a broad statute like Illinois' RTFA prevents a farm from being protected. A farm is a farm. It does not cease being a farm because of a change in ownership, or a temporary cessation due to foreclosure, or a change in the type of crop harvested. A change in local sensibilities can't make a farm no longer a farm under the Illinois RTFA.

Finally, the village argued that the RTFA *only* precludes making a farm a nuisance based on encroachment of nonagricultural use. Nothing in the statute prevents the village, absent those circumstances, from proclaiming a farm a nuisance "simply because it can." The village appears to be arguing the act only applies where the nuisance claim arises due to a change in the surrounding locale, which is the plain language of the statute, codifying the common law defense of "coming to the nuisance."

The court said that would lead to an absurd result where "the Act would bar enforcement of the ordinance if it was passed so that someone [could] build a house or two on a vacant lot across the road from the farm, but the Act would not protect the farm if no one wants to build a house on a vacant lot across the road from the farm."

Again, the court's linear logic is difficult to trace. The court, one supposes, is saying if no one is coming to the nuisance, then how can there be a nuisance? But as a dissenting judge notes, a change in the farm operation itself can give rise to a nuisance.

Regardless, this decision is important for a couple reasons. First, it treated a municipal ordinance banning farming as if it were a private nuisance action.

Second, the court refused to enforce a municipal ordinance despite there being no explicit language in the act purporting to nullify such ordinances. The court did so based on case precedent that preempts ordinances that infringe on the legislative intent of a state statute.

Finally, the judge applied broad protection to farms in a common sense, organic fashion resisting the urge to turn it into a bullet-point itemized list in a piece of code.

But that's also the weakness of the ruling, because, unlike common law analysis, when interpreting civil law statutes, if it's not written, it's not law.

The appellate court seems to rely heavily on the purpose of the act and on public policy as opposed to the plain language of the act.

The losing party appealed to the Illinois Supreme Court, but that court decided not to hear the case. It's not clear whether that was because the supreme court agreed with the appellate court or because the court was otherwise too busy. Like the U.S. Supreme Court, the Supreme Court of Illinois doesn't have to take every case.

Therefore, it's not clear whether this decision will hold up persuasively either in Illinois or in other jurisdictions.

But a farm-advocate can dream, can't he?

On a side note, the village ordinance was an expansion of an earlier ordinance that prohibited the keeping of farm animals or livestock within the village. One wonders if that ordinance too would be void under the RTFA had it been challenged.

If I were a town zoning board, my takeaway would be to draft a more targeted ordinance especially that could be characterized as a police power and not a "nuisance ordinance."

As a farmer, my takeaway is that, if you're facing the kind of ordinance described above, you might have a shot at challenging it in court. To put it in legal terms, you'd have a non-frivolous reason for challenging a local ordinance that conflicted with the RTFA, even if there wasn't explicit language in the RTFA purporting to void local ordinances. This is especially true in Dillon's Rule jurisdictions, or where there is a constitutional amendment, or where the RTFA itself contains a strongly-worded legislative purpose.[157]

---

[157] The strongest legislative purposes speak to the need to preserve agricultural land for reasons beyond the mere economic.

Indiana: Indiana Code 32-30-6-9[158]

Absent negligent operation, an agricultural operation or its appurtenances can't be a public or private nuisance due to any changed circumstances in the vicinity after it's been in operation for more than one year, provided there's no significant change in the type of operation and the operation wasn't a nuisance at the time it began.

The statute specifically states that a "significant change" does not include a change in the type of agricultural operation, a change in ownership or size, a change in participation in a government program, or adoption of new technology.

You lose protection if there's been a cessation of the operation of more than one year.

According to the Indiana Supreme Court, preventing an adjacent landowner from bringing a nuisance action does not constitute an unconstitutional taking.[159]

As to other court decisions, an appellate court held that although a hog operation isn't a nuisance *per se*,[160] a hog operation that begins five years *after* plaintiffs moved next door was not protected under the act.[161]

Also, the RTFA won't protect a farmer from a nuisance action arising out of a claim of *negligent* emptying of their dairy farm's manure pit.[162]

Farm defendants have the burden of proving they're protected under the act, but plaintiffs bringing the suit have the

---

[158] iga.in.gov/static-documents/8/f/b/4/8fb4ab51/TITLE32_AR30_ch6.pdf (static); codes.findlaw.com/in/title-32-property/in-code-sect-32-30-6-9.html.

[159] *Lindsey v. DeGroot*, 898 N.E.2d 1251 (Ind. Ct. App. 2009) (holding in opposite of the supreme court of Iowa; noting there's no case precedent in Indiana for holding a nuisance creates an easement (a negative use of the adjacent owner's land)).

[160] *Laux v. Chopin Land Assocs.*, 550 N.E.2d 100, 104 (Ind. Ct. App. 1990).

[161] *Wendt v. Kerkhof*, 594 N.E.2d 795 (Ind. Ct. App. 1992).

[162] *Stickdorn v. Zook*, 957 N.E.2d 1014 (Ind. Ct. App. 2011).

burden of proving negligence that would preempt an RTFA defense.[163]

As to changes and expansions, a change from grain farming to a hog raising operation could constitute a "significant change" in the type of agricultural operation; but an increase from 29 hogs to 300 hogs might not constitute a significant change.[164]

The court considering that case stated that farmers might be equitably estopped from relying on the RTFA if they knew the developer bought the adjacent land for residential purposes and yet proceeded with the hog farm anyway.[165]

However, if the developer bought the land but had done nothing with the land in more than a year, the developer was essentially "moving to the nuisance" one year after the hog farm had already been in operation.[166]

The hog farm lost. However, the trial court went too far in banning "any livestock raising operation" on the farm. The appellate court said the farm could conceivably raise some pigs (as it had been doing) without it being a nuisance.

Iowa: Iowa Code § 352.11[167]

Iowa was the first state to enact a Right to Farm statute. Oh, how the times have changed. Although the statute still exists, in reality, it is practically worthless.

Iowa's RTFA purports to give all farms or farm operations "located in an agricultural area" a break from nuisance regardless of when it started or expanded—full stop. Prevailing

---

[163] *Erbrich Products Co. v. Wills*, 509 N.E.2d 850 (Ind. Ct. App. 1987).

[164] *Laux v. Chopin Land Assocs.*, 550 N.E.2d 100, 103 (Ind. Ct. App. 1990).

[165] *Id.* at 104.

[166] *Id.* It wasn't enough that the developer's had paid a higher price for the land, or that his intention was to use the land for residences, or that he'd endured the loss of a sale because the developer hadn't built any residences, or platted a subdivision, or acquired a change in the agricultural zoning.

[167] coolice.legis.iowa.gov/Cool-ICE/default.asp?category=billinfo&service=IowaCode&ga=83&input=352.11.

defendants are entitled to receive fees and costs.

The county board of supervisors is responsible for designating "agricultural areas."

There are some exceptions to the RTFA for not following federal or state law; for operating in a negligent fashion; for damages created before the designation of the agricultural area; or, absent an act of God, for pollution or changes in waters of a stream, overflow of a person's land, or erosion.

But those exceptions and the act itself are largely moot. The Iowa Supreme Court objected to the blanket nuisance exception created for "agricultural areas." The supreme court held that a nuisance creates an easement on the affected property which, without compensation, would be an unconstitutional taking.[168]

However, once just compensation is paid for the easement, the purpose of Right to Farm Act can be maintained.

Practically, it looks like an injunction will not be issued against a farming operation but the farm will have to pay for the nuisance easement on surrounding land.

Other state courts, like Indiana above, have totally disagreed with the Iowa court's rationale.

Nevertheless, as the young people say, "It is what it is."

It is surprising that the Iowa legislature hasn't reformed its RTFA as have so many states when the courts make controversial holdings.

Finally, In Iowa, it's a criminal act in certain instances to make undercover recordings of an agricultural operation. For instance, it's a misdemeanor for an employee to "obtain access . . . under false pretenses [or] make a false statement or representation as part of an application [for employment] . . . with the intent of committing an act not authorized by the owner."

In addition to being a criminal act, the statute allows for a private civil cause of action.

---

[168] *Bormann v. Board of Sup'rs in & for Kossuth County*, 584 N.W.2d 309 (Iowa 1998).

Third-parties to the crime (i.e. conspirators) may also be liable both criminally and civilly. Hence, "agricultural operations targeted by an unlawful undercover investigation could seek to recover against the investigator, as well as their support and distribution networks, for lost profits that resulted from an investigation."[169]

It is important, then, for employers to make sure their policies about image and sound recording as well as data security are clearly communicated to their employees. If you're wondering what this looks like, you should probably consult a local attorney versed in employment law and contracts. But one could imagine employment manuals that include language limiting the use of personal electronic devices in certain areas or a complete ban. You'd also need to enforce these policies in a uniform manner which, yes, would mean getting on employees even for taking silly selfies. I know this goes against the grain of how we might like things to be, but this is the world we live in. You shouldn't assume animal rights groups and other social justice types are only interested in targeting CAFOs.

Kansas: Kansas Statutes Annotated § 2-3202[170]
Agricultural activities conducted on farmland, if consistent with good agricultural practices and established prior to surrounding agricultural or nonagricultural activities, are presumed to be reasonable and do not constitute a nuisance, public or private, unless the activity has a substantial adverse effect on the public health and safety.

To be protected, agricultural activities must follow "good agricultural practices." The statute does not define good practices.

[169] Daniel L. Sternberg, "Note: Why can't I know how the sausage is made?: How ag-gag statutes threaten animal welfare groups and the First Amendment," 13 Cardozo Pub. L. Pol'y & Ethics J. 625, 635 (2015).
[170] www.ksrevisor.org/statutes/chapters/ch02/002_032_0002.html.

However, if your activities conform with all laws, rules, and regulations at a local, state and federal level, the activity is legally presumed to be a good practice. In addition, it will be legally presumed that such activity does not affect the public health and safety.

Farmers can "reasonably" expand the scope of their agricultural activity including acreage, number of animals, or a change in the type of agricultural activity, so long as the activity complies with all applicable local, state and federal environmental codes, resolutions, laws, rules, and regulations.

A change in ownership is okay as is a temporary cessation or decrease in scope.

Feedlots are handled separately.[171] For feedlots, compliance with rules and regulations issued by the commissioner is "deemed to be prima facie evidence that a nuisance does not exist."[172]

Moreover, the law in Kansas limits applicable damages,[173] and provides for fees and costs where a farmer successfully defends a claim that he improperly used agricultural chemicals.[174]

Concerning reasonable changes and expansions, a 1993 case held that the RTFA did not protect a cow farm where it decided to confine 50 cows in 1,8 acre pen rather than giving them access to additional pasture. The court held the RTFA did *not* protect the farm since the nuisance was coming from the farm itself and not from moving to the farm and then complaining.[175] The residential home affected was a farm home in an agricultural area not an encroaching residential, nonagricultural use. Hence, a nuisance action can be brought, the RTFA notwithstanding, "if there has been a change in the nature of an agricultural activity on the property or if the operation does not conform with state

---

[171] Kan. Stat. Ann. § 47-1505 (2016).
[172] *Id.*
[173] Kan. Stat. Ann. § 2-3205 (2016).
[174] Kan. Stat. Ann. § 2-3204 (2016).
[175] *Finlay v. Finlay*, 856 P.2d 183 (Kan. App. 1993).

regulations."

Then again, a revision to the Kansas RTFA in 2013 added the language about farmers being able to "reasonably" change or expand the scope of activity. The preceding case is probably still good precedent, but it would depend on the facts of your case. Is the change or expansion "reasonable"? Does it comply with all rules and laws at the local, state, adnd federal level?

Finally, Kansas has a law making it a criminal act in certain instances to take undercover photos and videos in an animal facility with the intent to damage the enterprise.[176] In addition to being a criminal act, the statute allows for a private, civil cause of action.

Kentucky: Kentucky Revised Statutes § 413.072[177]

Agricultural and sivlicultural operations are protected from nuisance or trespass, public or private, and from violations of zoning ordinances that would "restrict the right of the operator of the agricultural or silvicultural operation to utilize normal and accepted practices." This protection is granted so long as it's been in operation for more than one year, wasn't a nuisance at the time it began, and the claim doesn't stem from negligent operation. Plus, the complained of nuisance has to be the result of changed conditions in or about the locality.

"Agricultural operations" is defined to include the usual list of animals and crops. But the definition is expanded to include "any generally accepted, reasonable, and prudent method for the operation of a farm to obtain a monetary profit that complies with applicable laws and administrative regulations, and is performed in a reasonable and prudent manner customary among farm operators." An agricultural operation can also include the

---

[176] Kan. Stat. Ann. § 47-1827 (2016).
[177] www.lrc.ky.gov/Statutes/statute.aspx?id=17860.

practice of sustainable agriculture, which is defined by the statute.

As to changes to the operation, the RTFA notes that silviculture operations can go a long time between harvests and are considered to be in continuous operation so long as the land is supporting an actual or developing forest.[178]

Meanwhile, you don't lose protected status by change of ownership or a cessation of less than five years (or a one year cessation after a government contract expires). Nor do you lose status by changing crops or methods of production due to the introduction of new and generally accepted technologies, unless the operation is substantially changed.

You aren't protected from suits based on damages on account of the pollution of the waters of any stream or ground water.

As for local rules and ordinances, Kentucky's RTFA and its counterpart in the zoning law statutes[179] has been called the "agricultural supremacy clause."[180]

These *explicitly* void local government ordinances that would make an agricultural operation a nuisance or seek abatement of it as a nuisance, trespass, or a zoning violation. For example, a county can't require approval before property can be divided into agricultural parcels.[181]

Though the RTFA provides great protection, if the nuisance is a result of changes made on your land or changes to your operation, you may lose protection under the RTFA.

---

[178] This can be very important. *See, Alpental Cmty. Club v. Gym. Soc'y*, 111 P.3d 257 (Wash. 2005) (Merely owning land with trees on it didn't constitute a forest activity. Hence, when a nuisance occurred while harvesting those trees, the tree farm wasn't covered under the RTFA because it hadn't met the one year requirement. The Washington state RTFA was immediately revised after this ruling to read more like Kentucky's.).

[179] Kentucky Rev. Stat. § 100.203(4) ("land which is used for agricultural purposes shall have no regulations" except for setbacks, dwellings, and to prevent flooding).

[180] *21st Century Dev. Co. LLC v. Watts*, 958 S.W.2d 25, 27 (Ky. Ct. App. 1997).

[181] *Nash v. Campbell County Fiscal Court*, 345 S.W.3d 811 (Ky. 2011).

For example, there was a class action case where property owners complained of a nuisance from the expansion of several local hog farming operations. Describing the smells using colorful language, the federal district court in Kentucky refused to dismiss the case under the RTFA because there was no evidence of changed conditions in the locality surrounding the barns, but rather, the landowners' residences were in existence when the new barns were built and began receiving hogs.[182]

If you are contemplating making a change to your operations, the question is whether it might be considered a "substantial change."

For further information about Right to Farm in Kentucky, the University of Kentucky offers an older, hand-typed but still perfectly cogent summary of the state right to farm law at the following   address: www.uky.edu/bae/sites/www.uky.edu.bae/files/AEU-65.pdf. It's worth reading even if only for a good trip down memory lane.

Louisiana: Louisiana Revised Code § 3:3603[183]

The civil code in Louisiana provides that persons engaged in agricultural operations in accordance with "generally accepted agricultural practices or traditional farm practices" are protected from nuisance suits brought by people who subsequently acquire land in the vicinity.

Agricultural operations include activities that are involved, directly and indirectly, in the production of food for human consumption or for animal food.

The exact language of the statute states, "no agricultural operation shall be deemed to be a nuisance in any action brought under [the civil code] or any other grant of authority authorizing

---

[182] *Powell v. Tosh*, 929 F. Supp. 2d 691 (W.D. Ky. 2013).
[183] legis.la.gov/Legis/Law.aspx?d=86319.

the suppression or regulation of public or private nuisances" provided the agricultural operation is conducted in accordance with generally accepted agricultural practices or traditional farm practices and where (1) the opponent acquired land after the date on which an agricultural operation was in existence; or (2) the operation was established prior to any change in the character of property in the vicinity; or (3) had existed for one year or more and the conditions deemed a nuisance have remained "substantially unchanged" since the operation began.

As for "generally accepted agricultural practices or traditional farm practices" those terms are defined by the code, and the code provides a legal presumption on behalf of farmers. The codes states, "Each person engaged in agricultural operations shall be presumed to be operating in accordance with generally accepted agricultural practices or traditional farm practices."[184]

Moreover, a defendant may recover fees and costs for frivolous lawsuits.[185]

Negligence, intentional injury or violation of state or federal laws or rules aren't protected under the RTFA.[186]

As for local ordinances, except for Jefferson Parish, no parish can adopt an ordinance that declares an agricultural operation to be a nuisance so long as it is following "generally accepted agricultural practices or traditional farm practices" and is not negligent. Neither may a parish adopt a zoning ordinance that forces the closure of any such agricultural operation. Moreover, municipal zoning ordinances don't apply when an agricultural operation is incorporated into the municipality by annexation.[187]

As for changes to the farm, the normal rotation of crops or livestock remains protected.

As a matter of interest, the Louisiana government, by statute,

[184] Louisiana Rev. Code § 3:3604 (LexisNexis 2016).
[185] Louisiana Rev. Code § 3:3605 (LexisNexis 2016).
[186] Louisiana Rev. Code § 3:3606 (LexisNexis 2016).
[187] Louisiana Rev. Code § 3:3607 (LexisNexis 2016).

seeks to avoid interfering in the use of or lowering the value of private agricultural property[188] and requires an impact assessment by the government before taking action that would lower the value of private agricultural property.[189] The law also sets forth owners' rights of action and remedies in such cases.[190]

Finally, it should be noted that there is a separate Right to Forest[191] provided for in the code as is the case in several states.

Despite the strong protection offered by Louisiana's RTFA, a farmer might still find herself in court! For example, an appellate court in Louisiana reversed a trial court's dismissal of a private nuisance suit against a produce business. The court said that, while the operation "might well be 'an agricultural operation conducted' 'in accordance with generally accepted agricultural practices,' *we find those to be issues which must be determined by the trier of fact.*"[192]

Just be aware that if you are sued in Louisiana, you may not be able to get the case dismissed (despite the fact that the code purports to give farmers a legal presumption). You may have to duke it out in court. And you can't get attorneys' fees and costs unless you can prove it was a frivolous suit, which is a very hard thing to do.

Maine: Maine Revised Statute Title 7, Section 153[193]
The Maine Agricultural Protection Act prevents a farm, farm operation, or agricultural composting operation from becoming a public or private nuisance where it is in compliance with state and federal laws, rules and regulations and (1) conforms to best management practices as determined by the commissioner; (2) follows rules relating to storage and use of farm nutrients; (3)

---

[188] Louisiana Rev. Code § 3:3608 (LexisNexis 2016).
[189] Louisiana Rev. Code § 3:3609 (LexisNexis 2016).
[190] Louisiana Rev. Code § 3:3610 (LexisNexis 2016).
[191] Louisiana Rev. Code § 3:3621 to :3624 (LexisNexis 2016).
[192] *Trosclair v. Matrana's Produce, Inc.,* 717 So. 2d 1257, 1259 (La. Ct. App. 1998).
[193] www.mainelegislature.org/legis/statutes/7/title7ch6sec0.html.

existed before the land within a one mile radius changed; and (4) hasn't materially changed the conditions or nature of its operation after a change in land use within that one mile radius.

If you're following best management practices, your agricultural practices can't be considered a violation of municipal ordinances.[194] Moreover, municipalities must provide a copy of proposed ordinances that affect farm operations to the state commissioner at least 90 days prior to enactment.[195]

Moreover, the RTFA provides a complaint resolution procedure.[196] The commissioner is empowered to investigate any complaints and to make determinations as to whether a farm is following best management practices.[197] If the commissioner finds that the problem is caused by use that doesn't conform to best management practices, he will determine the changes needed to comply, advise the person responsible for the farm, and subsequently determine if those changes are implemented. If a farm doesn't comply, the commissioner may initiate state action.[198]

Private actions against a farmer that are not brought in good faith or are frivolous or harassing may be sanctioned for attorneys' fees and costs.[199]

For additional information about Right to Farm in Maine, resources are available through Maine Farmland Trust.[200]

---

[194] 7 Me. Rev. Stat. Ann. § 154 (LexisNexis 2013).
[195] 7 Me. Rev. Stat. Ann. § 155 (LexisNexis 2013).
[196] 7 Me. Rev. Stat. Ann. § 156 (LexisNexis 2013).
[197] 7 Me. Rev. Stat. Ann. § 156 (LexisNexis 2013).
[198] 7 Me. Rev. Stat. Ann. § 157 (LexisNexis 2013).
[199] 7 Me. Rev. Stat. Ann. § 158 (LexisNexis 2013).
[200] www.mainefarmlandtrust.org/public-outreach-new/building-farm-friendly-communities/local-policies-ordinances.

Maryland: Maryland Code, Courts and Judicial Proceedings § 5-403[201]

The RTFA in Maryland provides agricultural, silviculture, or commercial fishing or seafood operations an affirmative defense[202] if they are sued for public or private nuisance provided (1) they have been underway for a year or more; (2) are compliant with federal, state and local health, environmental, zoning and permit requirements; and (3) aren't being conducted negligently. The RTFA also protects them from claims that they interfered with someone's use and enjoyment of their land.

Maryland is unique in that it requires all complaints to be mediated before they can be brought into court. The Maryland Department of Agriculture operates an agricultural conflict resolution center.[203] In addition, most localities have an agency authorized to manage agricultural disputes or, if not, the State Agricultural Mediation Program handles complaints. This administrative process must be followed. Administrative decisions can then be appealed to district court.

Maryland's RTFA doesn't prohibit the federal, state, or local government from enforcing health, environmental, zoning, or any other applicable law.

Specifically, the RFTA doesn't apply to any agricultural operation that is operating without a fully and demonstrably implemented nutrient management plan for nitrogen and phosphorus if otherwise required by law.

On a final note, every county in Maryland has also adopted a right to farm ordinance. This has, at times, caused confusion

---

[201] codes.findlaw.com/md/courts-and-judicial-proceedings/md-code-cts-and-jud-proc-sect-5-403.html.
[202] Paul Goeringer, "Maryland's Right-to-Farm Law," Maryland State Bar Association, Bar Bulletin, Apr. 15, 2016, www.msba.org/Bar_Bulletin/2016/04_-_April/Agriculture_Law__Maryland_s_Right-to-Farm_Law.aspx.
[203] mda.maryland.gov/Pages/acrs.aspx.

and led to inconsistent application of the law across counties.[204]

The University of Maryland's Department of Agriculture published an informative pamphlet on Right-to-Farm law.[205] They also produced a youtube video, which is available at: www.youtube.com/watch?v=7WY6klqmZWI.

Massachusetts: Massachusetts Annotated Laws chapter 243, section 6[206]

In one of the original right to farm cases, in 1963, a hog farm although "well and carefully operated" was shut down by a newly developed, residential neighborhood.[207]

The Right to Farm Act, passed in 1989, would prevent such nuisance actions against a farm or farming related operation for any "ordinary aspect" related to that operation. The farm has to have been in operation for more than a year to get protection. Negligent acts or actions "inconsistent with generally accepted agricultural practices" aren't covered.

It is not immediately clear who defines the term "generally accepted agricultural practice."

The Massachusetts statutes provide for procedural rules as to when and if the board of health finds a farming operation to be a nuisance. Except the board may not find odor and noise from generally accepted agricultural practices to be a nuisance.[208]

In addition, as many as ninety-eight towns have adopted right to farm bylaws.[209]

Moreover, Article 97 of the state constitution protects the peoples' "right to clean air and water, freedom from excessive

---

[204] Paul Goeringer & Lori Lynch, "Understanding Agricultural Liability: Maryland's Right-to-Farm Law," Univ. of Md. Cntr for Agric. and Nat. Resource Policy, Fact Sheet, July 2013, papers.ssrn.com/sol3/papers.cfm?abstract_id=2279619.
[205] papers.ssrn.com/sol3/papers.cfm?abstract_id=2279619.
[206] malegislature.gov/Laws/GeneralLaws/PartIII/TitleIII/Chapter243/Section6.
[207] *Pendoley v. Ferreira*, 345 Mass. 309 (1963).
[208] Mass. Ann. Laws 128 § 1A.
[209] www.mass.gov/eea/docs/agr/agcom/agcomm-map.pdf.

and unnecessary noise, and the natural, scenic, historic, and es-
thetic qualities of their environment" as well as the conservation,
development and utilization of agricultural and other natural resources. This amendment also allows the commonwealth to
acquire and maintain conservation easements.

Going back to the RTFA itself, an appellate court held that
construction of an "extraordinary" sand pile that expanded a
cranberry bog by an acre-and-a-half constituted a nuisance not
covered by the RTFA. The farmer took down 150 feet of trees,
which served as a buffer zone between the farm and the residents, and piled up 28-50 feet of sand less than 100 feet away
from the neighboring houses. According to the court, the old bog
was fine. The court even implied that some changes on a farm
won't invalidate RTFA protection. But such an "extraordinary"
change meant the new bog and sand wall constituted a nuisance.[210]

Michigan: Michigan Compiled Law Services § 286.473[211]

Michigan has one of the strongest Right to Farm acts in the country. However, it only protects *commercial* farming operations.
Moreover, small farms, homesteaders, and especially urban
farms may not be protected depending on their location in relation to residential areas.

The law specifically states that a commercial farm won't be a
public or private nuisance if it follows generally accepted agricultural and management practices (GAAMP) as determined by
the Michigan commission of agriculture. Nor will it be a nuisance provided that the farm existed before the change in
surrounding land use or occupancy of land within 1 mile of the

[210] *Francisco Cranberries LLC vs. Edward P. Gibney, Jr.*, 1999 Mass. App. Div. 223 (Mass.
App. Div. So. 1999).
[211] www.legislature.mi.gov/(S(z311dwua1q5dm5z1455iv4sv))/mileg.aspx?
page=getobject&objectName=mcl-286-473.

farm and wouldn't have been a nuisance before the change in land use or occupancy.[212]

The type of commercial production required for protection under the RTFA has been defined as "the act of producing or manufacturing an item intended to be marketed and sold at a profit."[213]

There is no one year timing requirement, and this is important. In one case, a hog farm was established two years *after* the plaintiff purchased her property. So it was almost as though the farm itself was coming to the nuisance. However, under the exact language of the law, an appellate court held that there had been no change from agricultural to residential usage within one mile of the farm and that the land use remained essential agricultural. Also, the hog farm complied with GAAMP. Therefore, it was protected under the RTFA.[214]

Michigan's RTFA follows the ALEC model act in protecting changes in ownership and size, temporary cessations or interruptions, enrolment in government programs, adoptions of new technology, and changes in the types of products being produced.

The act also allows prevailing defendant farmers to recover attorneys' fees and costs.[215]

Prior to the year 2000, many cases held that the RTFA did not protect famers from zoning ordinances.[216] An amendment to the RTFA, however, states that as of June 2000, local ordinances that

---

[212] It has been suggested that either of these two provisions may provide protection separately under the RTFA. Papedelis v. City of Troy, 2006 Mich. App. LEXIS 2748 *12 (Sept. 19, 2006) (unpublished non-binding decision, reversed on appeal on separate grounds) (noting "a business could conceivably move into an established residential neighborhood and start a farm or farm operation in contravention of local zoning ordinances as long as the farm or farm operation conforms to generally accepted agricultural and management practices.")

[213] *Shelby Charter Twp v Papesh*, 267 Mich. App. 92, 100-101 (2005).

[214] *Steffens v. Keeler*, 200 Mich. App. 179, 503 N.W.2d 675 (1993).

[215] Mich. Comp. L. Servs. § 286.473b (LexisNexis 2016).

[216] See, e.g., *City of Troy v. Papadelis*, 226 Mich. App. 90, 572 N.W.2d 246 (1997).

conflict with the RTFA or GAAMP are preempted.[217]

Hence, the RTFA preempted a local ordinance purporting to limit the raising of poultry on a parcel of land smaller than three acres.[218] Similarly, an ordinance precluding a pheasant hunting preserve was overturned.[219] A Michigan Appeals Court likewise held that a barn erected without a permit and in violation of zoning ordinances was yet protected because it was used for agricultural practices and conformed to generally accepted standards and was therefore protected under the RTFA.[220] Moreover, a court held that requiring a hog farm to obtain a special use permit might violate the RTFA.[221]

Still, the courts have sometimes upheld zoning restrictions. For instance, where an area was rezoned from agricultural to residential, and a hog farmer was subsequently denied a special use permit to continue farming, such did not violate the RTFA.[222] The court stated that the purpose of the RTFA was to provide immunity to agricultural operations from nuisance suits, rather than make farms totally exempt from zoning.

More recently, the Michigan Supreme Court held that a greenhouse and pole barn were not "incidental to the use for agricultural purposes of the land on which the building is located,"[223] which would exempt them from building code requirements. Nor did the ordinances as to permitting, size, height, bulk, floor area, construction or location of the buildings used for the greenhouse or related purposes conflict with any provision of the RTFA or GAAMPs. Hence, unlike the case of the barn mentioned above, the farm operation using the greenhouse and

---

[217] Mich. Comp. L. Servs. § 286.474(6) (LexisNexis 2016).

[218] *Charter Tp. of Shelby v. Papesh*, 267 Mich. App. 92 (Mich. Ct. App. 2005).

[219] *Milan Tp. v. Jaworski*, 2003 WL 22872141 (Mich. Ct. App. 2003), appeal denied, 470 Mich. 895, 683 N.W.2d 146 (2004).

[220] In *Northville Tp. v. Coyne*, 170 Mich. App. 446, 429 N.W.2d 185 (1988).

[221] *Belvidere Tp. v. Heinze*, 241 Mich. App. 324, 615 N.W.2d 250 (2000).

[222] *Padgett v. Mason County Zoning Com'n*, 2003 WL 22902878 (Mich. Ct. App. 2003).

[223] Mich. Comp. L. Servs. § 125.1502a(g) (LexisNexis 2016).

pole barn had to comply with local ordinances.[224]

A similar conclusion was reached by a Michigan Court of Appeals holding that a farm couldn't force a zoning board's hand on the permitting of or location of field driveways just because the driveway would make farming easier or more convenient. The ordinance regarding permittance on field driveways didn't conflict with the RTFA or GAAMP.[225]

In 2014, controversy arose [226] when the GAAMPs were amended to restrict protection of agricultural activities near residential areas.

The agriculture commission decided to revise the Livestock Site Selection GAAMP, adding Category 4 sites, which are locations that are primarily residential; that don't allow agricultural uses by right; and that are, according to the commission, not suitable for farm animals for purposes of the Right to Farm Act.

Under the Livestock Site Selection GAAMP, primarily residential areas are sites with more than 13 non-farm homes within an eighth of a mile of the livestock facility or one non-farm home within 250 feet of the livestock facility.

Because GAAMP protection is removed from such operations, local ordinances are no longer preempted.

This has left farming operations in urban areas at the mercy of local zoning boards especially in Muskegon and Brady Township. This use of GAAMP to contravene the express intent of the legislature in passing the RTFA is a bizarre yet typical bureaucratic move.

Consider the composition of the committee that rules on GAAMP in Michigan. The committee consists of "Recognized Professionals" only, including the following: conservation districts; industry representatives; Michigan department of environmental quality; professional consultants and contractors;

---

[224] *Papadelis v. City of Troy*, 478 Mich. 934 (Mich. 2007)
[225] *Scholma v. Ottawa County Rd. Comm'n*, 303 Mich. App. 12, 27 (Mich. Ct. App. 2013).
[226] growinggeorgia.com/features/2014/05/does-your-state-have-right-farm-protection.

professional engineers; USDA natural resources conservation service; and university agricultural engineers, and other university specialists.

Where are the small famers and other stakeholders on this list? How can a committee that purports to determine generally acceptable agricultural practices favor bureaucrats over actual farmers?

Finally, on a side note, while GAAMPs are supposed to encompass "the broadest possible sector of the state's agricultural industry," they "should not be construed as an exclusive list of acceptable practices."[227] Novel approaches that follow generally accepted agricultural management practices (GAAMP) may receive RTFA protection.

Ending on a complete tangent, a Michigan Court of Appeals held, without explanation, that the RTFA did not protect a dog kennel because it was not "akin to a farming operation."[228] Actually, we ran across several similar cases in other states.

Regarding the Right to Farm in Michigan, the Michigan Department of Agricultural and Rural Development maintains a helpful website[229] as does the Michigan Small Farm Council[230]. Small farmers wishing to start-up in Michigan are also encouraged to read the National Agricultural Center's "Michigan Direct Farm Business Guide."[231]

Minnesota: Minnesota Statutes Annotated § 561.19[232]

An agricultural operation (pertaining to crops, livestock, poultry,

---

[227] *Almont Tp. v. Dome*, 1997 WL 33354480 (Mich. Ct. App. 1997)(mobile home on tree farming operation held a generally accepted practice).
[228] *Township of Groveland v. Bowren*, 1998 WL 1988929 (Mich. Ct. App. 1998).
[229] www.michigan.gov/mdard/0,4610,7-125-1599_1605--,00.html.
[230] www.michigansmallfarmcouncil.org.
[231] A. Bryan Endres, Lisa R. Schlessinger, & Rachel Armstrong. "Michigan Direct Farm Business Guide," nationalaglawcenter.org/wp-content/uploads//assets/articles/MIdirectfarm.pdf (2015).
[232] www.revisor.mn.gov/statutes/?id=561.19.

dairy products or poultry products) that has been in operation for more than 2 years shall not become a private or public nuisance *as a matter of law* provided (1) it is located in an agriculturally zoned area; (2) that it complies with federal, state, or county laws, regulations, rules, and ordinances and any permits; (3) and operates according to generally accepted agricultural practices.

"Generally accepted agricultural practices" means those practices commonly used by other farmers in the county or a contiguous county in which a nuisance claim arises.

An expansion or significant alteration can reset the 2 year clock as to that expansion or alteration. Expansion means at least a 25 percent increase in the number of animals. Significant alteration does not mean a change in ownership between family members, or a temporary cessation or interruption in cropping activities, or adoption of new technologies, or a change in the crop produced. That being said, under an earlier version of the RTFA, an appeals court held that changing a cow-calf operation to a CAFO may constitute a significant alteration.[233]

Even before the 2 years runs, agricultural operations can get a rebuttable presumption that they aren't a public or private nuisance.

However, the RTFA doesn't apply to an animal feedlot facility with a swine capacity of 1,000 or cattle capacity of 2,500. Nor does it apply to actions by a public authority to abate a particular condition that is a public nuisance, nor to any enforcement action brought by a local unit of government related to zoning.

For example, the Minnesota Court of Appeals held that setback requirements for a feedlot operation are a "valid exercise of land use regulatory authority by a township."[234] The feedlot's expansion had been approved by the state air pollution agency.

---

[233] *Haas v. Tellijohn*, 1996 Minn. App. LEXIS 289 (Minn. Ct. App. Mar. 12, 1996).
[234] *Canadian Connection v. New Prairie Tp.*, 581 N.W.2d 391 (Minn. Ct. App. 1998).

However, the township's zoning ordinance didn't focus on pollution control, rather it bore a rational relationship between the regulating the feedlot and the odor concern of residents. Hence, the denial of a variance and conditional use permit was upheld by the trial court and on appeal.[235]

In 2004, the legislature amended the language of the RTFA in reaction to a negative appellate court ruling. At that time, the legislature imposed a two-year limitation on nuisance suits and barred nuisance claims against operations that are conducted according to "generally accepted agricultural practices" as a matter of law, thereby creating an affirmative defense.[236] The legislature also added a rebuttable presumption for the first two years of operation as well.

Minnesota does not allow neighbors to characterize an odor nuisance complaint as a trespass.[237] However, chemical pesticide drift can constitute a trespass.[238] In addition, unlike the Iowa Supreme Court, a federal district court in Minnesota held that the RTFA is not unconstitutional and doesn't create an easement over neighboring property.[239]

Mississippi: Mississippi Code Annotated § 95-3-29[240]

Proof that an agricultural operation has existed for one year or more is an absolute defense to public or private nuisance actions if the operation is in compliance with all applicable state and federal permits. Agricultural operations should also use its

---

[235] *Id.*
[236] *Wendinger v. Forst Farms, Inc.,* 662 N.W.2d 546, 550 (Minn. Ct. App. 2003).
[237] *Id.*
[238] *Johnson v. Paynesville Farmers Union Coop. Oil Co.,* 802 N.W.2d 383 (Minn. Ct. App. 2011).
[239] *Overgaard v. Rock County Bd. of Comm'rs,* 2003 U.S. Dist. LEXIS 13001 (D. Minn. July 25, 2003).
[240] law.justia.com/codes/mississippi/2013/title-95/chapter-3/section-95-3-29.

materials, machines, and structures in accordance with best agricultural management practices.

Although no part of the RTFA voids local ordinances, other portions of the statute limit municipalities' ability to require permits for agricultural land [241] and farm buildings [242]. However, according to the state attorney general, such exemptions do not include farm residences. [243]

The courts in Mississippi have held that a wood processing plant was an agricultural operation protected by the RTFA [244] as was a paper mill [245] and a cotton gin [246]. Similarly, hog farms operating for more than a year were protected by the RTFA absent a "substantial change" in the operation. [247] (A 2009 amendment to the statute later removed the "substantial change" language in favor or requiring compliance with state and federal permits, thereby making the statutory protection even stronger.) Moreover, that court found no private cause of action in the air and water pollution control law. [248] But a public nuisance claim for water pollution under the Mississippi Air and Water Pollution Control Law could proceed. [249]

---

[241] Miss. Code Ann. § 17-1-3 ("[N]o permits shall be required with reference to land used for agricultural purposes, including forestry activities as defined in Section 95-3-29(2)(c), or for the erection, maintenance, repair or extension of farm buildings or farm structures, including forestry buildings and structures, outside the corporate limits of municipalities.")

[242] Miss. Code Ann. § 17-2-7 (prohibiting the enforcement of building codes, including the International Building Codes, on farm structures) & 19-5-9 ("codes shall not apply to the erection, maintenance, repair or extension of farm buildings or farm structures" with certain exceptions).

[243] *Cummings*, Sept. 29, 2006, A.G. Op. 06-0436.

[244] *Hill v. Koppers, Inc.*, 2010 U.S. Dist. LEXIS 5036 (N.D. Miss. Jan. 20, 2010).

[245] *Leaf River Forest Prods., Inc. v. Ferguson*, 662 So. 2d 648 (Miss. 1995).

[246] *Bowen v. Flaherty*, 601 So. 2d 860 (Miss. 1992).

[247] *Norman v. Prestage Farms, Inc.*, 2007 U.S. Dist. LEXIS 24456 (N.D. Miss. Mar. 30, 2007).

[248] *Norman v. Prestage Farms, Inc.*, 2007 U.S. Dist. LEXIS 24456 (N.D. Miss. Mar. 30, 2007); see also *In re Moore*, 310 B.R. 795 (Bankr. N.D. Miss. 2004).

[249] *Leaf River Forest Prods., Inc. v. Ferguson*, 662 So. 2d 648 (Miss. 1995).

Missouri: Missouri Revised Statutes § 537.295[250]

In 2014, Missouri enshrined right to farm in its constitution.[251] Missouri also has a fairly strong RTFA. It states that agricultural operations are immune from public or private nuisance actions based on any changed conditions in the locality, provided they've been in operation for more than a year and weren't a nuisance at the time the operation began.

The RTFA doesn't apply if the alleged nuisance is the result of negligent or improper operation. Neither does it preclude actions based on injuries from water pollution or overflow of land.

The RTFA does not appear to apply to agricultural operations located within the limits of any city, town, or village. However, according to Professor Erin Morrow Hawley, "Some of the local zoning laws might be subject to challenge under the constitutional amendment."[252]

As to changes, the RTFA allows for "reasonable expansions" provided all county, state, and federal environmental codes, laws, or regulations are met. Additionally, such expansions shouldn't create a substantially adverse effect upon the environment or be a hazard to public health or safety or create a measurably significant difference in environmental pressures on neighbors because of increased pollution. Reasonable expansions can be of like kind but aren't meant to include complete

---

[250] www.moga.mo.gov/mostatutes/stathtml/53700002951.HTML.

[251] Missouri Const. Art. I § 35. "That agriculture which provides food, energy, health benefits, and security is the foundation and stabilizing force of Missouri's economy. To protect this vital sector of Missouri's economy, the right of farmers and ranchers to engage in farming and ranching practices shall be forever guaranteed in this state, subject to duly authorized powers, if any, conferred by article VI of the Constitution of Missouri."

[252] Marshall Griffin, "Is Missouri ballot measure boon for family farms or just big corporations?," St. Louis Public Radio, July 17, 2014 at 3:22 PM EDT, www.pbs.org/newshour/rundown/would-missouri-ballot-measure-benefit-family-farms-or-corporations. See also the discussion on Illinois. See also the discussion in the section on Illinois regarding the case *Vill. of Lafayette v. Brown*, 27 N.E.3d 687 (Ill. App. 2015)(holding RTFA voided local ordinance despite containing no language to that effect), appeal denied *Vill. of Lafayette v. Brown*, 392 Ill. Dec. 370 (Ill. 2015).

relocations. And expansions of livestock and poultry operations should ensure they can handle waste properly.

Once an operation obtains protected status, that status can be assigned, sold, and inherited. It is not waived by temporary cessation or diminutions in the size of the operation.

Finally, a defendant farm can recover attorney fees for frivolous lawsuits.

A Missouri statute also limits damages, precluding recovery of non-economic damages caused by a nuisance that emanates from property used primarily for agriculture.[253] This limitation was found to be perfectly constitutional.[254]

The right to farm statute hasn't produced a published or unpublished case. Probably because of negotiated settlements of the sort that led farmers to seek a constitutional amendment.

In Missouri, it may be criminal in certain instances to enter an agriculture facility to damage it, or free animals, or even to make undercover recordings.[255] Regarding surreptitious recordings, it is illegal to "[o]btain[] access to an animal facility by false pretenses for the purpose of performing acts not authorized by the facility." In addition to being a criminal act, this statute allows for a private civil cause of action including attorney fees and costs. Hence, it is important for farms to have clear policies regarding what is and isn't authorized when it comes to photos, videos, and other recordings in non-public areas of their facilities and to follow up on those policies consistently (yes, even as regards selfies posted to social media accounts).

---

[253] Mo. Rev. Stat. § 537.296 (2016).
[254] *Labrayere v. Bohr Farms, LLC*, 2015 Mo. LEXIS 29 (Apr. 14, 2015).
[255] Mo. Rev. Stat. § 578.405 (2016).

Montana: Montana Code Annotated § 27-30-101(3)[256] & 45-8-111(4)[257]

The constitution of the state of Montana requires the Montana Legislature to "protect, enhance, and develop all of agriculture."[258] Hence, Montana statutes contain incentives to keep agricultural land in production. Meanwhile, local communities are somewhat restricted when it comes to agricultural mitigation.

In fact, according to the legislature's express purpose: "In order to sustain Montana's valuable farm economy and land bases associated with it, farmers and ranchers must be encouraged and have the right to stay in farming. It is therefore the intent of the legislature to protect agricultural activities from governmental zoning and nuisance ordinances."[259]

The definition of agriculture is quite broad and includes selling produce marketed at roadside stands or farm markets.[260] In fact, the words "agriculture" or "agricultural" are used as many as 200 times in the Montana code.[261] Many counties have also passed right to farm ordinances.[262]

Moreover, local zoning ordinances can't reach agricultural activities outside their boundaries, nor may they prohibit agricultural activities that were established outside the corporate

---

[256] leg.mt.gov/bills/mca/27/30/27-30-101.htm.
[257] leg.mt.gov/bills/mca/45/8/45-8-111.htm.
[258] Mont. Const. Art. 8 § 1.
[259] Mont. Code Ann. § 76-1-901 (2016).
[260] Mont. Code Ann. § 76-1-902 (2016).
[261] "Agricultural Protection in Montana: Local Planning, Regulation, and Incentives," at n. 62, Land Use Clinic, University of Montana School of Law (Spring 2012), *available at* www.farmlandinfo.org/sites/default/files/AgReport09.10.12_UMontana_LawSchool.pdf.
[262] Terri Adams, "The Right To Farm May Be Eroded if Producers Do Not Know Their Rights," The Prairie Star, January 28, 2013, 12:30 am, www.theprairiestar.com/news/regional/the-right-to-farm-may-be-eroded-if-producers-do/article_be160648-6660-11e2-84f8-0019bb2963f4.html (Musselshell, Petroleum, Wheatland, Meagher, Golden Valley, Lincoln, Prairie, Ravali, Beaverhead and Madison counties had RTF ordinances at the time of publication in 2013 and other counties were considering the same).

limits of the municipality and later incorporated into the municipality by annexation.[263]

According to a report issued by the Montana School of Law's Land Use Clinic, "Ultimately, local governments must work with agricultural operators to identify the most significant incentives to keep agricultural lands in production."[264]

In Montana, a farming operation doesn't become a public[265] or private[266] nuisance because of its normal operation as a result of a change in surrounding use provided that the farm has been in operation longer than the complaining resident has been in possession or longer than a neighboring commercial establishment has been in operation.

Perhaps due to the comprehensiveness of Montana's right to farm laws, we're aware of no cases that have interpreted the RTFA. However, based on similar language in other jurisdictions, one could foresee challenges based on a significant change to the farm operation itself (rather than neighbors "coming to the nuisance") as well as challenges as to what constitutes "normal operations."

Finally, in Montana, it's a criminal act in certain instances to make undercover recordings at an animal facility with the intent to damage the enterprise and with the intent to commit criminal defamation. The intent requirement narrows the applicability of the law quite a bit.[267] In addition to being a criminal act, the statute allows for a private, civil cause of action.

---

[263] Mont. Code Ann. § 76-1-903 (2016).
[264] "Agricultural Protection in Montana: Local Planning, Regulation, and Incentives," Land Use Clinic, University of Montana School of Law (Spring 2012), available at www.farmlandinfo.org/sites/default/files/AgReport09.10.12_UMontana_LawSchool.pdf.
[265] Mont. Code Ann. § 45-8-111 (2016).
[266] Mont. Code Ann. § 27-30-101(3) (2016).
[267] Mont. Code Ann. § 81-30-103(2) (2016).

Nebraska: Nebraska Revised Statute §2-4403[268] & 81-1506[269]

In 2016, an amendment to the state constitution was proposed but was stopped by lawmakers before it could reach the voters.[270]

Certainly, Right to Farm in not as staunchly protected in Nebraska as in other big ag states.

For instance, there seems to be no check on zoning ordinances, no affirmative defense for following generally accepted agricultural management practices (GAAMP), no defense for reasonable changes in operation, no limitation on the types of damages that may be awarded (i.e. compensatory damages versus injunctive relief), no recovery of attorneys' fees and court costs, and no one-year statute of repose.

It's no wonder farmers sought an amendment to the state constitution. However, one might suggest that a few carefully crafted statutory changes as catalogued in the paragraph above might produce the desired effect.

Still, the state RTFA does provide protection from public or private nuisance actions where the farm existed before a change in the land use of the surrounding land and would not have been a nuisance before the change.[271]

Moreover, an animal feeding operation isn't a nuisance provided "reasonable techniques" are employed to keep dust, noise, insects, and odor at a minimum; provided the feedlot is in compliances with zoning regulations; and provided the complaint is brought by someone who came into possession of land subsequent to a permit being issued for the feedlot or subsequent to

---

[268] nebraskalegislature.gov/laws/statutes.php?statute=2-4403.
[269] nebraskalegislature.gov/laws/statutes.php?statute=81-1506.
[270] Joanne Young, "Right-to-farm resolution pulled from debate," Journal Star, Mar. 24, 2016, journalstar.com/legislature/right-to-farm-resolution-pulled-from-debate/article_40c9dff6-c62a-53d9-8fe9-4acd6a8bb005.html.
[271] Neb. Rev. Stat. § 2-4403 (2016).

the operation of a feedlot for which, based on inspection, no permit is required.[272]

The courts have made it clear that the RTFA protects farms only where the changes have been to "in and about the locality" of such farm or farm operation. Hence, in two separate cases, suits against hog farming operations were not barred by the RTFA. In one case, the changes giving rise to the nuisance had been changes on the hog farm itself.[273] In another case, there had been no change in and about the locality of the hog farm because the complaining neighbor had resided there prior to the commencement of the farming operation.[274]

Nevada: Nevada Revised Statute Annotated 40.140(2)[275]

The Nevada RTFA takes a bit of a novel approach, providing a legal presumption "[t]hat an agricultural activity conducted on farmland, consistent with good agricultural practice and established before surrounding nonagricultural activities is reasonable."

To restate, there is a legal presumption that farm operations following good agricultural practices are not a "nuisance" as a matter of law. Note also that there is no one or two year timing requirement as there is in other states.

The statute goes on to say that agricultural activity as described "does not constitute a nuisance unless the activity has a substantial adverse effect on the public health or safety." The term "substantial" sets the bar high and may not include normal odors.

Moreover, an agricultural activity that doesn't violate a federal, state or local law, ordinance or regulation is presumed to

---

[272] Neb. Rev. Stat. § 81-1506 (2016).
[273] *Flansburgh v. Coffey*, 370 N.W.2d 127 (Neb. 1985).
[274] *Cline v. Franklin Pork, Inc.*, 361 N.W.2d 566 (Neb. 1985).
[275] www.leg.state.nv.us/NRs/NRS-040.html#NRS040Sec140.

constitute good agricultural practice. As one local law firm notes, "The reverse is not necessarily true, though. The existence of a violation of law or ordinance does not necessarily equate with nuisance, as the law is written."[276]

No statutory language was found regarding preemption of local ordinances. Nor was any case law found that addresses or interprets Nevada's RTFA.

New Hampshire: New Hampshire Revised Statute § 432:33[277]
The New Hampshire RTFA is short and sweet. It says that an agricultural operation can't be a public or private nuisance as a result of changed conditions in the surrounding area provided the operation has been in operation for one year and wasn't a nuisance when it began. However, the RTFA doesn't apply if the operation is "injurious to public health or safety."

Also, the RTFA doesn't protect farmers from nuisance actions brought for negligent or improper operation. That being said, agricultural operations *shall not* be found to be negligent or improper if they conform to federal, state and local laws, rules and regulations.[278]

The RTFA, notwithstanding, local health officers retain the authority to regulate the prevention and removal of nuisances.[279] Such authority is not limitless and has been circumscribed by statute.

For example, the New Hampshire Supreme Court upheld a zoning ordinance that prevented a Christmas tree farm from holding weddings.[280] In response, the legislature amended the

---

[276] "Nevada's Pro-Farmer Right to Farm Law," posted on June 25, 2013, ballard-lawpractice.com/nevadas-pro-farmer-right-to-farm-law.
[277] www.gencourt.state.nh.us/rsa/html/XL/432/432-33.htm.
[278] N.H. Rev. Stat. Ann. § 432:34 (2016).
[279] N.H. Rev. Stat. Ann. § 147:1 (2016).
[280] *Forster v. Town of Henniker*, 167 N.H. 745 (N.H. 2015).

RTFA, adding language to protect agritourism.[281] Presumably, weddings are now protected (though it's still not entirely clear as we'll explain below).

After the supreme court case, the legislature specifically stated, "*Agricultural activities and agritourism* shall not be unreasonably limited by use of municipal planning and zoning powers or by the *unreasonable interpretation* of such powers."[282]

Unreasonable interpretations include: "the failure of local land use authorities to recognize that *agriculture and agritourism* as defined when practiced in accordance with applicable laws and regulations, are traditional, fundamental and accessory uses of land throughout New Hampshire, and that a prohibition upon these uses cannot necessarily be inferred from the failure of an ordinance or regulation to address them."[283]

Similarly, statutory law creates a legal presumption that, "whenever agricultural activities are not explicitly addressed with respect to any zoning district or location, they shall be deemed to be permitted there."

Looking at the preceding sections, one legal expert summed it up this way: "[I]t still remains a question whether weddings on farms are now *per se* permitted under the new agritourism amendment. The amendment doesn't actually say that, but towns certainly have the ability to explicitly include weddings on farms as a permitted agritourism/agricultural use. However, if towns do not include that permission in their zoning ordinance, question remains about whether weddings (and other activities) must always be allowed."[284]

Moving back to the RTFA, making changes to your agricultural activity will not cause you to lose protection. The statute

---

[281] 2016 NH Ch. 267.
[282] N.H. Rev. Stat. Ann. § 672:1(III)(b) (2016).
[283] N.H. Rev. Stat. Ann. § 672:1(III)(d) (2016).
[284] E-mail from Attorney Margaret Byrnes of New Hampshire Municipal Association, on file with the author, Nov. 2016.

says that qualifying agricultural uses may expand without restriction or be altered to meet changing technology or markets and may even be changed to another agricultural use provided such expansions or alterations comply with all federal and state laws, regulations, and rules including best management practices adopted by the commissioner of agriculture, markets, and food.[285]

However, newly established or reestablished operations as well as "significant expansions" (including farm retail operations) may be subject to special exception, building permit, or other local land use board approval.

Although the RTFA seems simple in New Hampshire, you can see how the interaction between the RTFA and zoning ordinances as well as court decisions can complicate matters. Standing by to assist you is FTCLDF as well as other farm organizations and governmental entities that can help you understand your rights in New Hampshire.[286]

New Jersey: New Jersey Statute Annotated § 4:1C-10[287]

A *commercial* agricultural operation, activity, or structure gets an irrebuttable presumption that the operation doesn't constitute a public or private nuisance (or any other invasion or interference with the use and enjoyment of another's land or property) provided it meets three requirements. First, the operation must conform to agricultural management practices passed by the state agricultural development committee's formal rule making process; *or* it must be determined to constitute a generally accepted agricultural operation or practice on a case-by-case basis by the county ag board or, where no board exists, by the state committee. Second, the operation has to conform to federal or state statutes, rules or regulations. Finally, the operation mustn't

---

[285] N.H. Rev. Stat. Ann. § 674:32(b) (2016).
[286] www.nh.gov/oep/resource-library/agriculture/index.htm.
[287] www.nj.gov/agriculture/sadc/rtfprogram/rtfact.

pose a direct threat to public health and safety (as determined by the state agricultural development committee).

To be a commercial farm, an operation has to prove up profits: $2,500 for larger farms (at least five acres of contiguous or non-contiguous land), or $50,000 for small farms, or $10,000 for bee keeping operations.[288]

In addition, for farms established after 1998, the farm has to be located in an area in which agriculture is a permitted use under the municipal zoning ordinance and is consistent with the municipal master plan.

As to preemption of local ordinances, the state code protects commercial agricultural activities[289] "[n]otwithstanding the provisions of any municipal or county ordinance, resolution, or regulation to the contrary," provided they follow "generally accepted agricultural operation or practice" as determined by the state agricultural development committee.[290]

The state code and subsequent court decisions indicate that any complaints (even those relating to zoning violations) must first go through the county agriculture development board or state agricultural development committee where no board exists.[291] The committee should determine whether the operation complained of complies with a published agricultural management practices or with a generally accepted agricultural operation or practice. In so doing, the board must also consider the impact of municipal land use ordinances as well as the impact of various municipal concerns like water run-off, stream encroachment, and the creation of impervious areas. Committee

---

[288] N.J. Stat. Ann. § 4:1C-3 (2016).
[289] N.J. Stat. Ann. § 4:1C-9 (2016).
[290] *Township of Franklin v. Hollander*, 338 N.J. Super. 373, 389 (N.J. App. Div. 2001).
[291] This requirement is similar to Maryland above.

decisions are final unless appealed.[292]

Again, per the statute, the courts have indicated that the board, rather than a trial court, has primary jurisdiction to determine whether generally acceptable agricultural practices have been met. For example, it was up to the board to determine whether a farmer's placement of trailers along a property line[293] was an acceptable agricultural practice or whether an organic farm's interference with the flow of water[294] was a nuisance.

Finally, it bears repeating that the New Jersey RTFA applies only to commercial farms. For instance, a horse breeding operation was held not to be a commercial farm so as to be eligible for local zoning preemption where the landowner failed to meet the $2,500 annual gross profits requirement.[295] In that case, the farmer couldn't prove up future sales.[296]

The state has published a handy guidebook[297] as well as a factsheet[298] on the Right to Farm in New Jersey.

New Mexico: New Mexico Statutes Annotated § 47-9-3[299]

New Mexico's Right to Farm act states that an agricultural operation or facility (including operation of a roadside farm market[300]) doesn't become a private or public nuisance based on

---

[292] *Township of Franklin v. Hollander*, 338 N.J. Super. 373, 392 (N.J. App. Div. 2001); *id.* at 375 (N.J. App. Div. 2001) (noting that the following decision had been superseded by statute *Villari v. Zoning Bd. of Adjustment of Deptford*, 649 A.2d 98 (N.J. App. Div. 1994)).

[293] *Curzi v. Raub*, 999 A.2d 1182 (N.J. Super. App. Div. 2010).

[294] *Borough Of Closter v. Abram Demaree Homestead, Inc.*, 839 A.2d 110 (NJ App. Div. 2004), *certification denied*, 845 A.2d 1254 (NJ 2004).

[295] N.J. Stat. Ann. 4:1C-3a (2016).

[296] *In re Tavalario*, 901 A.2d 963 (NJ App. Div. 2006).

[297] www.nj.gov/agriculture/sadc/rtfprogram/resources/guidebook.pdf.

[298] www.nj.gov/agriculture/sadc/rtfprogram/resources/factsheet.pdf.

[299] No free, updated statute seemed currently available online. Here is an alternative with links to the old statute and the amendment: Tiffany Dowell, "New Mexico Amends Right to Farm Act," Texas A&M Agrilife Extension, Mar. 14, 2016, agrilife.org/texasaglaw/2016/03/14/new-mexico-amends-right-to-farm-act.

[300] N.M. Stat. Ann. § 47-9-5(11) (2016).

any changed condition in or about the locality, provided the operation has been ongoing for more than a year and wasn't a nuisance at the time the operation began.

The one year limit on lawsuits begins to toll when an agricultural operation commences or an agricultural facility is built. Subsequent expansion or the adoption of new technology does not change the established date of operation.

The usual caveats apply: the RTFA won't protect negligent or illegal operations nor will it prevent lawsuits based on pollution or changed conditions of waters or streams, or based on overflow.[301]

The RTFA says that local ordinances or resolutions that would make an agricultural operation or facility a nuisance (or provide for abatement as if it were a nuisance) shall not apply when an agricultural operation is located within the corporate limits of any municipality.

Meanwhile, following a string of lawsuits against dairy farmers state-wide,[302] an amendment in March 2016 added that prevents one who buys, leases, rents, or occupies property adjacent to a previously established agricultural operation or facility from bringing a nuisance suit unless there has been a *substantial* change in the nature and scope of the operations.

Finally, under the RTFA attorneys' fees and costs may be awarded to a farming defendant for frivolous lawsuits.[303]

There doesn't appear to be any reported cases interpreting New Mexico's RTFA.

---

[301] N.M. Stat. Ann. § 47-9-6 (2016).
[302] Emily Crowe, "Ag producers, advocates want decades old Right to Farm Act," Clovis News Journal, July 25, 2013, cnjonline.com/2013/07/25/ag-producers-advocates-want-decades-old-right-to-farm-act-updated.
[303] N.M. Stat. Ann. § 47-9-7 (2016).

New York: N.Y. Agric. & Mkts § 308[304]

In the late 1960s, New York's Constitution was amended to protect agricultural land.[305] Part of that protection now includes the Right to Farm Act.[306]

New York's RTFA is a bit unique. It protects *commercial* agricultural operations that follow "sound" agricultural practices (as determined by the Commissioner of Agriculture and Markets). These operations must take place on land designated as part of an agricultural district,[307] or land that is used in agricultural production that is subject to assessment by the Commissioner.

A "sound" agricultural practice is one that is "necessary for the on-farm production, preparation and marketing of agricultural commodities."[308] Whether an agricultural practice is sound is determined on a case by cases basis.[309] Decisions are subject to court review, but the commissioner's decisions must be given great deference by the courts.[310]

Local governments cannot unreasonably regulate farm operations in an agricultural district unless they can show that the

---

[304] law.justia.com/codes/new-york/2010/agm/article-25-aa/308.

[305] N.Y. Constitution, Art. 14, § 4.

[306] N.Y. Agric. & Mkts § 300 et seq.

[307] "There are 8.6 million acres of land in agricultural districts encompassing about 26,000 farms in New York State." Presentation by Dr. Bob Somers, New York State Agriculture and Markets, Aug. 13, 2014, Dryden Neptune Fire Hall, available at dryden.ny.us/wp-content/uploads/2014/10/August-13th.pdf.

[308] N.Y. Agric. & Mkts § 308(1)(b) (2016).

[309] A New York Court held that a case-by-case evaluative approach to determining which generally accepted agricultural practices are protected under the RTFA did not constitute a breach of a complainant's constitutional due process rights contra Iowa's *Bormann*, 584 N.W.2d 309. Rather, the court deferred to the agricultural commission's authority, expertise, and experience. *Pure Air & Water, Inc. v. Davidsen*, 668 N.Y.S.2d 248 (N.Y. App. Div. 1998) (hog farm's manure management program requested and obtained advisory opinion thereby preempting suit under RTFA).

[310] The standard of review is "arbitrary and capricious." *Pure Air and Water Inc. v. Davidsen*, 668 N.Y.S.2d 248 (N.Y. App. Div. 1998).

farm operation is a threat to public health and safety.[311] However, the commissioner *may* consult a number of entities in making the determination of whether an agricultural practice is "sound," including the municipality in which the agricultural practice being evaluated is located. If a landowner feels a local rule is unreasonably restrictive, she can request a review from the department. The Ag Protection board's opinion, then, is binding as to how a local rule affects a farm operation.

Take, for example, the case where a goat farm was able to reopen a vacant barn in the center of town turning it into a microcreamery.[312] The farmer and her husband had been engaged in a long-running, contentious "dialogue" with the local planning board; however, as it turns out, due to the RTFA, the board had no say in whether farming can occur on the property. Rather, the board could only approve a special use permit.[313] According to one planning board member, "If it's agriculture, it's none of our business."[314] Affected neighbors could take sanitation complaints to the public health agency and pollution and runoff issues to the city department of environmental protection.

According to Kristin Janke Schneider, an environmental planner at New York's Delaware County Planning Department, the right to farm law "protects the right to farm on any parcel where it can be conducted safely."[315]

---

[311] N.Y. Agric. & Mkt. § 305; N.Y. Town Law § 283-a (2016). Presentation by Dr. Bob Somers, New York State Agriculture and Markets, Aug. 13, 2014, Dryden Neptune Fire Hall, available at dryden.ny.us/wp-content/uploads/2014/10/August-13th.pdf.

[312] Baylen Linnekin, "Right-to-Farm Debate Heats Up Controversies over laws in all 50 states that protect the rights of farmers to actually farm," Oct. 24, 2015, reason.com/archives/2015/10/24/right-to-farm-debate-heats-up.

[313] Julia Reischel, "Andes goat farm receives permit for creamery in center of town," Watershed Post, Oct. 15, 2015, www.watershedpost.com/2015/andes-goat-farm-receives-permit-creamery-center-town.

[314] *Id.*

[315] *Id.*

That being said, farms must still comply with applicable zoning laws. For example, even though agriculture buildings are exempt from all of the codes including electrical, plumbing, and fire; nevertheless, approval has to be received from the health department for there to be a bathroom on site.[316] Similarly, a stand that people buy produce from is exempt, but if the public has to enter the stand, it is not exempt.[317]

Regarding zoning and farm-related activities. Non-farm related activities are not covered by the RTFA; however, activities may be covered where the intent is to sell or market products grown by the farm:[318] for example, a café where all the food served was produced on the farm itself or a wedding where the wine served exceeded the rental fee for the facility and the food was catered. In contrast, the sale of sandwiches, ice cream, and drinks at farm markets is not covered.[319]

Moving back to the RTFA itself, which protects "sound" practices, farmers have the burden of proving their practices are "sound."[320] To do so, a farmer can request a determination by the commission. Such a determination that the operation is following "sound" practices would trigger protection under the RTFA. Commission opinions are made available to the public.[321] Farmers can rely on these decisions only in a very general sense; every decision is made on a case by case basis and your operations may differ somewhat.

Getting into the nitty-gritty, state law precludes an action for private nuisance where agricultural activities were commenced

---

[316] Presentation by Dr. Bob Somers, New York State Agriculture and Markets, Aug. 13, 2014, Dryden Neptune Fire Hall, available at dryden.ny.us/wp-content/uploads/2014/10/August-13th.pdf.
[317] Id.
[318] Id.
[319] Id.
[320] Concerned Area Residents for the Environment v. Southview Farm, 834 F. Supp. 1410 (W.D. N.Y. 1993).
[321] Opinions available linked at the following page: www.agriculture.ny.gov/ap/agservices/agdistricts.html.

prior to the surrounding activities, have not increased substantially in magnitude or intensity, and have not been determined to be the cause of conditions dangerous to life or health.[322]

Also, there are some of the usual limitations on RTFA protection. First, it doesn't prevent an action for personal injury or wrongful death. Neither does it prevent a public nuisance action even for "sound" practices that creates a harm to the health, safety, or welfare of the general public. Finally, other causes of action might still apply including trespass and violation of other state and federal environmental laws such as water pollution or land overflow.[323]

Of particular concern, manure and bacteria from manure may be considered a "pollutant" such that they aren't covered by RTFA protection or standard farm insurance.[324] For instance, even if spreading manure over one's farm is considered "sound," if subsequent heavy rains causes bacteria to leach into a neighboring well, the farmer might yet be liable despite the RTFA. You may want to make sure your insurance covers such contingencies.

Finally, attorney fees and costs shall be awarded to defendant if the agricultural practice was sound according to an opinion issued by the commissioner prior to the start of any trial or settlement of such action.[325]

---

[322] N.Y. Pub. Health § 1300-c (2016).

[323] *See, e.g., Concerned Area Residents for the Environment v. Southview Farm*, 843 F. Supp. 1410 (W.D.N.Y. 1993).

[324] David Ganje, "Full Immunity From Liability Not Possible For New York Farmers," www.lexenergy.net/full-immunity-from-liability-not-possible-for-new-york-farmers. *See e.g. Concerned Area Residents v. Southview Farm*, 834 F. Supp. 1410 (W.D.N.Y. 1993); *Space v. Farm Family Mut. Ins. Co.*, 235 A.D.2d 797, 798-99 (3d Dep't 1997).

[325] N.Y. Agric. & Mkt § 308-a (2016).

North Carolina: North Carolina General Statute § 106-701[326]

North Carolina, the second largest producer of hogs, has been a hotbed of complaints and lawsuits in the recent years. In July 2013, 588 private complaints were filed against Smithfield hog operations[327] as well as state environmental actions and even a complaint with the EPA's office of civil rights for "environmental racism."[328]

Although Smithfield is owned by Chinese interests, local farmers who grow hogs for them have been driven into bankruptcy by the actions.[329]

The lawsuits alleging odor and other nuisances may run afoul of the RTFA. However, now is a good time to remind you that being protected by the RTFA won't necessarily keep you from being sued.

The RTFA in North Carolina is pretty standard:

> No agricultural or forestry operation or any of its appurtenances shall be or become a nuisance, private or public, by any changed conditions in or about the locality outside of the operation after the operation has been in operation for more than one year, when such operation was not a nuisance at the time the operation began.

Agricultural operations are limited to "any facility for the production for *commercial* purposes of crops, livestock, poultry, livestock products, or poultry products."

[326] www.ncga.state.nc.us/EnactedLegislation/Statutes/HTML/ByArticle/Chapter_106/Article_57.html.
[327] David Bennett, "Right to Farm laws being tweaked across nation: Know your rights as a farmer," Aug 7, 2013, Delta Farm Press, deltafarmpress.com/government/right-farm-laws-being-tweaked-across-nation.
[328] Lisa Sorg, "NC DEQ asks EPA to dismiss civil rights complaint over hog farms," The Progressive Pulse, Sept.9, 2016, pulse.ncpolicywatch.org/2016/09/09/nc-deq-asks-epa-to-dismiss-civil-rights-complaint-over-hog-farms.
[329] Elisse Ramey & Gina DiPietro, "HOG FARMERS: Ongoing lawsuit has pushed us to bankruptcy" WITN.com, July 29, 2015, www.witn.com/home/headlines/HOG-FARMERS-Ongoing-lawsuit-has-pushed-us-to-bankruptcy-319701941.html.

Fundamental changes to agricultural operations aren't protected. For example, converting a turkey farm into a high volume commercial swine facility constituted a fundamental change that wasn't protected by the RTFA.[330] However, that case took place before a 2013 amendment (which may have been adopted to curtail the aforementioned lawsuits) stating that fundamental changes do not include a change in ownership, size, or type of product produced; an interruption of three years or less; participation in a government program; or the introduction of new technology. Because the shift from turkeys to hogs might be a change it type, it is unclear whether the aforementioned case is still good law.

Plaintiffs might argue, however, that the change was to the farm itself and not a changed condition "in or about the locality outside of the operation." For example, in another case, a youth camp in operation prior to the commencement of a hog farm weren't barred from suing the hog farm in nuisance because there had been no change in the locality, rather the change was to the farm itself.[331]

Regardless, the RTFA does not apply to negligent or improper operations. Nor does it cover injuries or damages stemming from pollution or change in the condition of the waters of any stream, or on account of overflow of lands.

The RTFA preempts local ordinances that would make an operation a nuisance or provide an abatement thereof, provided the nuisance isn't a result of negligent or improper operation.

Moreover, the county zoning enabling act exempts "bona fide farms" from county zoning,[332] but there is no similar exception for city zoning. And cities do regulate farms (even claiming

---

[330] *Durham v. Britt*, 451 S.E.2d 1 (NC Ct. App. 1994).
[331] *Mayes v. Tabor*, 77 N.C. App. 197, 334 S.E.2d 489 (1985).
[332] N.C. Gen. Stat. § 153A-340 (2016).

extraterritorial jurisdiction) to the extent possible including using general police power ordinances limiting the number of
animals that may be kept or preventing CAFOs from locating
within one mile of the city.[333]

As for RTFA lawsuits, attorney fees and costs shall be
awarded to the defendant for frivolous or malicious action or to
the plaintiff if the farmer asserts a frivolous or malicious affirmative defense.

On a side note, the North Carolina Farm Act of 2013 limits
the liability of farmers selling to wholesellers and resellers (not
direct to consumers) where they've complied with the USDA
Good Agricultural Practices and Good Handling Practices and
other food safety requirements.[334]

That same act also limits liability for the inherent risks of
agritourism, like petting zoos and animal-based educational
programs.[335]

And finally, in January 2016, North Carolina put a law on the
books, overriding the governor's veto, that makes it illegal for an
employee to surreptitiously record (make undercover recordings of images or data) in person or by planting a camera in non-
public areas.[336] The law criminalizes other activities as well such
as intentionally entering and damaging private property. And
the law provides a civil cause of action for you to recover. Un-
surprisingly, the constitutionality of that law is being
challenged.[337]

According to the language of that new law, third parties to

---

[333] "Intensive Livestock Operations In North Carolina: Cases And Materials" *in* Environmental and Conservation Law, No. 2, Mar. 1996 published by the Institute of
Government. The University of North Carolina at Chapel Hill, available at
wfs.enr.state.nc.us/pages/WL_right_to_farm_law.htmL.
[334] North Carolina law 2013-265 s638, available at ncleg.net/Ses-
sions/2013/Bills/Senate/HTML/S638v7.html.
[335] *Id.*
[336] N.C. Gen. Stat. § 99A-2 (2016).
[337] Animal Legal Defense Fund, "Taking Ag-gag to Court," available at aldf.org/
cases-campaigns/features/taking-ag-gag-to-court (last visited Oct. 25, 2016).

the crime (i.e. conspirators) may also be liable both criminally and civilly. Hence, you may be able to recover for any damages against the investigator, as well as their support and distribution networks. Regardless, this highlights the importance of having a clearly-stated employee policy as to what is or is not allowed concerning electronic devices. Don't assume animal rights advocates are only interested in CAFOs.

North Dakota: North Dakota Century Code § 42-04-02[338]

North Dakota became the first state to enact a constitutional amendment enshrining the right "to engage in modern farming and ranching practices" that "employ agricultural technology, modern livestock production and ranching practices."[339] Although the amendment is strongly worded, the RTFA itself is nothing to write home about.

In North Dakota, agricultural operations cannot be a private or public nuisance based on any changed conditions in surrounding land use after they've been in operation for more than a year. However, negligent or improper operation isn't protected. So, if you accidentally let your hogs escape more than twenty times, you're going to have a bad time.[340] Nor does the RTFA apply to pollution or any changed condition in the waters of any stream, or on account of any overflow of lands.[341]

The RTFA voids any local ordinances or resolutions that make agricultural operations a nuisance (or provide for an abatement thereof) unless the nuisance resulted from negligence or improper operation or if the operation was inside city limits prior to the enactment of the RTFA.

---

[338] www.legis.nd.gov/cencode/t42c04.html.
[339] N.D. Constitution, Art. 11.
[340] *State v. Hafner*, 587 N.W.2d 177 (N.D. 1998).
[341] N.D. Cent. Code § 42-04-03 (2016).

North Dakota was also one of the first states to make it criminal in certain instances to enter an agriculture facility to make
undercover recordings.[342] In addition to being a criminal act, the
statute allows for a private, civil cause of action.

Ohio: Ohio Revised Code Annotated § 929.04[343] & 3767.13[344]

Just to switch things up a bit, Ohio has two statutes that provide
immunity to a nuisance action with regard to agricultural activities. The first exemption, which we'll get to below, pertains to
activities located in an agricultural district.[345] The second provides a cause of action in nuisance for noxious, noisome, or
offensive smelling animals.[346] *Except*, immunity is conferred onto
persons who are (1) operating agriculture-related activities, (2)
outside a municipal corporation, (3) in accordance with generally accepted agricultural practices, and (4) in such a manner so
as not to have a substantial, adverse effect on the public health,
safety, or welfare. If so, not only are such persons exempt from
a lawsuit under this section, they're also exempt from any similar ordinances, resolutions rules or other enactments of a state
agency or subdivision including those that prohibit excessive
noise.[347]

So, if a nuisance suit is brought against you based on that
particular statute (including a nuisance action brought by state
officials pursuant to a statute, regulation, or ordinance), you
might be protected under that particular provision.[348] For instance, a township board of trustees couldn't enjoin an animal
shelter because it was operated by state license "in accordance

[342] N.D. Cent. Code § 12.1-21.1-02 (2016).
[343] codes.ohio.gov/orc/929.04v1.
[344] codes.ohio.gov/orc/3767.13.
[345] Ohio Rev. Code Ann. § 929.04 (2016).
[346] Ohio Rev. Code Ann. § 3767.13 (2016).
[347] Ohio Rev. Code Ann. § 3767.13 (2016).
[348] *See Winkelmann v. Cekada*, 738 N.E.2d 397 (9th Dist. Medina County 1999).

with generally accepted agricultural practices."[349]

But wait, there's more. There is also a common-law cause of action in nuisance that can apply even if you would otherwise qualify for immunity under the above exception, because that section creates an exception *only* to the cause of action named in that statute. It doesn't create an exception to a common-law nuisance cause of action.[350]

However, as mentioned briefly above, there is a statute that provides a complete defense[351] for *all* civil nuisance actions for agricultural activities *within an agricultural district* provided (1) the activities were established before the plaintiff's activity or acquired interest, provided (2) the plaintiff isn't a farmer, and provided (3) the activities don't conflict with state, federal or local laws relating to the alleged nuisance and were conducted in accordance with generally accepted practices.

As one appellate court noted, "[T]he extension of this complete defense to qualified entities is essentially a legislative determination that conduct falling within the purview of the statute, regardless of its frequency or intensity, is not a nuisance as a matter of law."[352] Also, once you've been exonerated one time under this statute, you can't be sued again by the same plaintiff for the same activities.[353]

Farmers can apply to have their land designated as an agricultural district provided the farm is at least ten acres or, if less, that certain minimum profits were made in the past three

---

[349] *Bd. of Brimfield Twp. Trs. v. Bush*, 2007 Ohio 4960 (Ohio Ct. App., Portage County Sept. 21, 2007).

[350] *Id.*; *Moody v. Wiza*, 2007-Ohio-5356, NaN-P68 (Ohio Ct. App., Ottawa County Oct. 5, 2007).

[351] Technically, the statute provides an affirmative defense that must be raised in your Answer. It is waived if not properly raised. *Eulrich v. Weaver Bros.*, 165 Ohio App. 3d 313 (Ohio Ct. App., Logan County 2005).

[352] *Harmon v. Adams*, 2002-Ohio-2103 (Ohio App. 3d Dist. 2002).

[353] *Id.* (second suit barred by the legal doctrine of *Res Judicata*).

years.[354] If the farm is located within a municipality, the farmer also has apply with the city or town. If the locality tries to modify or reject an application, it must demonstrate a substantial adverse impact on the provision of municipal services, the efficient use of land, orderly growth and development, or the public health, safety, or welfare.[355]

For agricultural districts that are later annexed into a municipality, the municipality cannot review the classification unless certain exceptions are met (such as selling or transferring the property to a person outside the family, or having voted for or signed onto the annexation).[356]

Ohio statutes also prevent townships from prohibiting the use of land for agricultural purposes,[357] limits a township's ability to regulate agriculture in platted subdivisions,[358] and limits the regulation of farm markets.[359]

Hence, the Supreme Court of Ohio recently held a township had no business regulating a winery out of business even though only 5% of the grapes used in the wine were produced on site.[360] The township argued that the property should be characterized as a restaurant or retail business, which isn't allowed in residential areas.[361] The trial court and the appellate court agreed that the primary use of the property wasn't agricultural, growing grapes, but making and selling wine; therefore it didn't qualify for the exception to the zoning of land for agricultural purposes.[362] However, the Supreme Court held, "There is no

[354] "Ohio's Right to Farm Law," Ohio Farm Bureau Federation: Public Policy Department, Jan. 2009, available at www.portagefb.org/wp/wp-content/uploads/2011/02/Ohios-Right-to-Farm-Brochure.pdf
[355] Id.
[356] Id.
[357] R.C. 519.21(A) (2016).
[358] R.C. 519.21(B) (2016).
[359] R.C. 519.21(C) (2016).
[360] Terry v. Sperry, 130 Ohio St. 3d 125 (Ohio 2011).
[361] Id.
[362] Id.

requirement . . . that the vinting and selling of wine be a second-ary or subordinate use of the property or that viticulture be the primary use of the property. A township may not prohibit the use of a property for vinting and selling wine if any part of the property is used for viticulture."[363]

In fact, to further protect agritourism and wineries, a 2016 amendment was passed that specifically includes protections for agritourism and explicitly states that the zoning exemptions in-clude "buildings or structures that are used primarily for vinting and selling wine and that are located on land any part of which is used for viticulture."[364]

However, exemption from municipal codes and regulations doesn't apply to medical marijuana cultivators, processors, or retail dispensaries.[365] Bummer, dudes.

Oklahoma: Oklahoma Statute Annotated Chapter 50, Section 1.1[366]

Oklahoma voters rejected a Constitutional amendment enshrin-ing Right to Farm.[367]

Nevertheless, in Oklahoma, agricultural activities (*including improvements and expansions*) are yet protected from nuisance lawsuits by the RTFA. The statute reads:

> Agricultural activities conducted on farm or ranch land, if con-sistent with good agricultural practices and established prior to nearby nonagricultural activities, are presumed to be reasonable and do not constitute a nuisance unless the activity has a substantial ad-verse affect (sic) on the public health and safety.

---

[363] *Id.* at 131.
[364] 2015 Ohio SB 75.
[365] 2016 Ohio HB 523.
[366] webserver1.lsb.state.ok.us/OK_Statutes/CompleteTitles/os50.rtf.
[367] Joe Wertz & Logan Layden, "Oklahoma Divided: How Geography Influenced the Vote on 'Right-to-Farm'," NPR: State Impact, Nov. 10, 2016, stateimpact.npr.org/oklahoma/2016/11/10/oklahoma-divided-how-geography-influenced-the-vote-on-right-to-farm.

If that agricultural activity is undertaken in conformity with federal, state and local laws and regulations, it is presumed to be good
agricultural practice and not adversely affecting the public health
and safety.

A 2009 amendment[368] brought improvements and expansions, including non-contiguous expansions, within the
definition of "agricultural activities."

That amendment also put two-year time limit[369] on nuisance
actions against agricultural activities. No nuisance action can be
brought against agricultural activities that have been in operation for two years or more prior to the date of the lawsuit.
Subsequent expansions of physical facilities or the adoption of
new technologies don't reset the two-year clock.

That same amendment added language awarding attorney
fees and costs to defendants for frivolous lawsuits.

However, the RTFA doesn't preclude the application of state
and federal laws like the Clean Water Act or the Oklahoma Concentrated Animal Feeding Operations Act and the Oklahoma
Registered Poultry Feeding Operations Act.

Oklahoma's RTFA contains no language specifically precluding zoning and other ordinances and regulations. But
municipalities are prevented by statute from issuing zoning regulations affecting "the erection or use of the usual farm buildings
for agricultural purposes, the planting of agricultural crops or
the extraction of minerals."[370]

---

[368] 2009 OK. ALS 147.
[369] Using technical legal jargon, many courts have found the time limit to be a Statute
of Repose.
[370] 19 Okl. Stat § 868.11.

Oregon: Oregon Revised Statute § 30.936[371]

Oregon's RTFA is fairly broad, conferring immunity from private actions or claims for relief in nuisance or trespass for farming practices on land zoned for farm use (irrespective of when such use began). RTFA protection does not apply to damage to commercial agricultural products or death or serious physical injury.[372] Moreover, immunity applies "regardless of whether the farming or forest practice has undergone any change or interruption."[373]

The RTFA confers similar immunity on farming practices that are allowed as a preexisting nonconforming use (for example, that were in place before a contrary zoning ordinance was passed or before a farm was annexed). The farming practice must have existed before the conflicting nonfarm use that gave rise to the nuisance or trespass claim and provided the farming practice hasn't significantly increased in size or intensity since November 1993 or the date in which the urban growth boundary changed to include the farming practice within its limits.[374]

Nuisance and trespass actions that are excluded by law include but aren't limited "actions or claims based on noise, vibration, odors, smoke, dust, mist from irrigation, use of pesticides and use of crop production substances."[375]

Farming practice is defined as "a mode of operation on a farm that: (a) Is or may be used on a farm of a similar nature; (b) Is a generally accepted, reasonable and prudent method for the operation of the farm to obtain a profit in money; (c) Is or may become a generally accepted, reasonable and prudent method in conjunction with farm use; (d) Complies with applicable laws;

---

[371] www.oregonlaws.org/ors/30.936.
[372] Or. Rev. Stat. § 30.936 (2016).
[373] Id.
[374] Or. Rev. Stat. § 30.937 (2016).
[375] Or. Rev. Stat. § 30.932 (2016).

and (e) Is done in a reasonable and prudent manner."[376]

Moreover, local government ordinances that make farming practices a nuisance or trespass (or provide for an abatement) are invalid.[377] Based on this section, a farmer who used a special breed of dog to herd goats and to bark at predators was exempt from a local noise nuisance ordinance because the dogs were engaged in a farming practice. The government didn't prove the dogs were not being used in a generally accepted, reasonable, and prudent manner.[378]

Finally, the statute states "the prevailing party *shall* be entitled to judgment for reasonable attorney fees and costs incurred at trial and on appeal."[379]

It has been suggested that Oregon's RTFA is unconstitutional, likely due to its breadth. Arguments against Oregon's RTFA are similar to those that were used to find the Iowa RTFA unconstitutional.[380] However, as we've noted above, several other state and federal courts have rejected the Iowa court's reasoning.[381] But, honestly, the statutes in those states weren't as sweepingly broad and their courts not as progressive. So far, however, such constitutional challenges haven't gone anywhere in Oregon. Potential plaintiffs may have been dissuaded by the possibility of having to pay attorney fees and costs if they lose.[382]

---

[376] Or. Rev. Stat. § 30.930(2) (2016).

[377] Or. Rev. Stat. § 30.935 (2016).

[378] *Hood River County v. Mazzara*, 89 P.3d 1195 (Or. App. 2004).

[379] Or. Rev. Stat. § 30.938 (2016).

[380] Lisa N. Thomas, "Forgiving Nuisance and Trespass: Is Oregon's Right-to-Farm Law Constitutional?" 16 J. Envtl. L. & Litig. 445 (Fall 2001).

[381] *See n. 29 supra.*

[382] Mateusz Perkowski, "Opponents Of Oregon's 'Right To Farm' Law Can Revive Lawsuit," Capital Press, Jan. 14, 2015, www.capitalpress.com/Oregon/20150114/opponents-of-oregons-right-to-farm-law-can-revive-lawsuit.

Pennsylvania: 3 Pennsylvania Statutes §§ 951 to 957[383]

Pennsylvania, like other states, adopted its RTFA in order to limit "the circumstances under which agricultural operations may be the subject matter of nuisance suits and ordinances."[384]

To that end, the Pennsylvania RTFA precludes nuisance actions against agricultural operations lawfully operating for a year or more prior to the filing of the action,[385] provided also that the conditions or circumstances giving rise to the nuisance claim have existed substantially unchanged since the established date of operation and are normal agricultural operations.

If there's been a substantial expansion or alteration to the physical facilities or operations, those are protected if the substantial alteration or expansion has been extant for one year prior to the suit or has been addressed in and is in compliance with a nutrient management plan prior to the changes. Hence there is a one-year "reset" for substantial changes after which they are protected under the RTFA.[386]

To qualify as a "normal agricultural operation" the farm has to be engaged in certain prescribed activities and be *ten or more contiguous acres* or have an anticipated yearly gross *income of at least $10,000*.

Agricultural operations also include any new activities, practices, equipment and procedures consistent with technological development within the agricultural industry, as well as the operation of the usual sort of farm machinery.[387]

Whether an operation qualifies as a "normal agricultural operation" is a broad inquiry, "focusing on the practice in general,

---

[383] codes.findlaw.com/pa/title-3-ps-agriculture/pa-st-sect-3-954.html.
[384] 3 Penn. Stat. § 951 (2016).
[385] This language likely makes it a statute of repose. *See Trinity River Auth. v. URS Consultants*, 889 S.W.2d 259, 261 (Tex. 1994).
[386] *Id.* at 19, n. 18.
[387] 3 Penn. Stat. § 952 (2016).

not on whether the defendant in this particular instance conducted the practice in accordance with accepted industry
standards and regulations."[388] For instance, "the manner in
which biosolids are applied at a particular site is not determina-
tive of the practice's normalcy," but, rather, "go[es] to the
underlying merits of the nuisance claim."[389]

Although whether a practice is a "normal agricultural operation" may be a fact intensive inquiry, it's not to sort that goes to
a jury but rather, in order to preserve legal certainty, uniformity,
and consistency, should be resolved as a matter of application of
law to fact by a judge.[390]

This is an interesting twist because at least two other state
courts found that such determination cannot be made by the
judge but must be made by the trier of fact.[391]

Nevertheless, according to the Pennsylvania court, because
the legislature intended to limit nuisance suits, "the definition of
'normal agricultural operation' is read expansively, taking into
account new developments in the farming industry."[392]

The RTFA doesn't preclude the state from protecting the
public health, safety, and welfare nor does it preclude a municipality from enforcing state law. Nor does it protect operations

---

[388] *Gilbert v. Synagro Cent., LLC*, 131 A.3d 1, 20 (Pa. 2015).
[389] *Id.* at 23.
[390] *Id.* at 18, holding:

> [W]hether an activity is a "normal agricultural operation"—is a categorical
> inquiry for the court. Otherwise, agricultural practices would be subject to
> nuisance suits based on varying local perceptions of what constitutes a
> "normal agricultural operation," as parochial opinion differs from jury to
> jury and juror to juror. What is common in one area may be foreign to an-
> other. Having courts apply the RTFA's definitions achieves the meaningful
> degree of legal certainty, uniformity, and consistency that the RTFA was in-
> tended to provide to farms.

[391] *Pestey v. Cushman*, 1994 Conn. Super. LEXIS 3275, 6-7 (Conn. Super. Ct. Dec. 15,
1994); *Trosclair v. Matrana's Produce, Inc.*, 717 So. 2d 1257, 1259 (La. Ct. App. 1998).
[392] *Id.* at 20.

that are conducted in violation of applicable federal, state, or local statutes or government regulations[393]; or pollution or harm to
waters or flooding of lands[394].

A recent Pennsylvania Supreme Court case illustrates the
tensions created by precluding nuisance actions under the
RTFA.[395] The supreme court recently held against a group of
long-time farm residents in their nuisance suit against a farm
that applied biosolids (aka sewage-sludge and residential sep-
tage) as fertilizer.

The farm's conservation plan required it to use no-till agriculture, which provides many environmental benefits, but
prevented them from plowing the biosolids under the soil,
which is a major odor control strategy. The trial court, after extensive discovery, granted summary judgment in favor of the
farm, holding that the use of biosolids did not constitute a "substantial change" that might negate RTFA protection. The
appellate court disagreed.

However, the Supreme Court held reversed the appellate
court decision. Holding in favor of the farm, the supreme court
said that the use of biosolids fell within "normal agricultural operation."[396] Again, it's not a matter of whether that farm applied
biosolids correctly but whether the use of biosolids is normal
practice, which it is.

The dissent, however, argued that the no-till method might
not be a normal practice.

Interestingly, as noted above, the supreme court said the
RTFA is the type of statute a judge may decide as a matter of law
without a jury.[397]

---

[393] 3 Penn. Stat. § 954 (2016).
[394] 3 Penn. Stat. § 955 (2016).
[395] *Gilbert v. Synagro Cent., LLC*, 131 A.3d 1, 2 (Pa. 2015).
[396] *Id.* at 2.
[397] *Id.*

Moving from case law into zoning ordinances, local municipalities may not define a normal agricultural operation as a nuisance provided the operation doesn't have a direct adverse effect on the public health and safety.[398] Moreover, direct on-site sales must be allowed, regardless of any ordinances, public nuisance or zoning prohibitions, provided the on-site sales are made by a landowner who produces 50% or more of the commodities sold (unless crop failure prevents the landowner from reaching the 50% limit).[399]

In addition, Section 603(h) of the Pennsylvania Municipalities Planning Code expressly states that "nothing in this subsection shall require a municipality to adopt a zoning ordinance that violates or exceeds the provisions of the...the [Right to Farm Law]." And the Agricultural, Communities and Rural Environment Act[400] prohibits municipalities from adopting ordinances that prohibit or limit "normal agricultural operations" and directs the attorney general to review local ordinances at a landowner's request.

Hence the attorney general successfully challenged an ordinance that purported to limit "intensive agriculture" as being arbitrary, vague, unreasonable and inviting discriminatory enforcement. In that instance, a 1500-foot setback in the ordinance was preempted by the provisions of the Nutrient Management Act, and the ordinance was found to unreasonably restrict farm structures and farm practices (by requiring daily removal of waste from a poultry operation), in violation of the Agricultural Area Security Law.[401]

In other cases, zoning ordinances held up against a poultry

---

[398] 3 Pa. Stat. § 953 (2016).
[399] *Id.*
[400] 3 Pa. Stat. §§ 311-318 (2016).
[401] *Com., Off. of Atty. Gen. ex rel. Corbett v. Richmond Tp.*, 360 MD 2006, 2010 WL 3155927 (Pa. Cmmw. 2010).

slaughterhouse[402] and denial of a building permit and variances for an equipment storage facility on a one-acre parcel of land.[403] The courts found that the RTFA protects an operation from enforcement of "nuisance ordinances as they apply to any normal agricultural operation" but the code "provides an array of 'zoning purposes' from which [the courts] conclude[d] that the power to zone is in no way limited to suppression of nuisances."[404] These rulings are in keeping with similar decisions in other jurisdictions that distinguish nuisance ordinances from other types of ordinances, for example, ordinances that are a valid exercise of a municipality's police power.

As usual, when it comes to fighting zoning ordinances, your mileage may vary.

Rhode Island: Rhode Island General Laws § 2-23-5[405]

The RTFA in Rhode Island appears to be quite weak. It only protects agricultural operations found to be a public or private nuisance based on the following itemized list: (1) odor from livestock, manure, fertilizer, or feed as occasioned by generally accepted farming procedures; or (2) noise from livestock or equipment used in normal, generally accepted farming procedures; or (3) dust created during plowing or cultivation operations; or (3) the use of pesticides, rodenticides, insecticides, herbicides, or fungicides.

The RTFA doesn't cover agricultural operations conducted in a malicious or negligent manner or operations that violate federal or state law as to the use of pesticides, rodenticides, insecticides, herbicides, or fungicides.[406]

---

[402] *Wellington Farms, Inc. v. Township of Silver Spring*, 679 A.2d 267 (Pa. Commw. Ct. 1996).

[403] *In re Appeal of Burger*, 17 Pa. D. & C.4th 280, 1992 WL 573037 (C.P. 1992).

[404] *Id.*

[405] webserver.rilin.state.ri.us/Statutes/TITLE2/2-23/INDEX.HTM.

[406] R.I. Gen. Laws § 2-23-4 (2016).

In addition, the RTFA only applies to actions seeking to abate (stop) a nuisance.[407]

"Agricultural operations" is defined as *commercial* enterprises of the usual sort and also includes other uses such as viniculture, the production of fiber, furbearing animals, and bees.[408]

In 2014, an amendment was added recognizing the acceptability of certain agritourism activities including, but not limited to, "the display of antique vehicles and equipment, retail sales, tours, classes, petting, feeding and viewing of animals, hay rides, crop mazes, festivals and other special events" as important ways to preserve agriculture.[409] Weddings are not yet listed, though a bill was introduced to add weddings in the 2015 legislatives session.[410] Nevertheless, weddings may be considered sufficiently similar to the above list to warrant inclusion under the statute.

Presumably the inclusion of agritourism activities was meant to shield them from nuisance complaints; however, the language of the statute does not seem sufficiently clear. For instance, nuisance protection is only extended to the numbered list above. It's not clear how the agritourism activities, despite their inclusion in the definitions section of the RTFA, fit into the enumerated list of protected agricultural operations. Furthermore, the statute goes on to say the RTFA doesn't apply to nonagricultural activities, uses, or operations.

That being said, the inclusion of agritourism under the definition of agricultural operations might protect such from local nuisance ordinances[411] (although a municipality might issue

[407] R.I. Gen. Laws § 10-1-1 (2016).
[408] R.I. Gen. Laws § 2-23-6 (2016).
[409] 2014 R.I. ALS 406.
[410] Tracee M. Herbaugh, "R.I. House panel to mull bill adding weddings to farm uses," Providence Journal, May 27, 2015, www.providencejournal.com/article/20150527/NEWS/150529326.
[411] R.I. Gen. Laws § 23-19.2-1 (2016).

other regulations like alcohol sales, hours of operation, and police presence at the event). And, inclusion in the statute might
qualify agritourism activities as generally accepted farming
practices[412] for what that's worth.

I don't envy the court that first faces the difficulty of untwin-
ing Rhode Island's RTFA with regards to agritourism, or the
farm that relies on the RTFA for protection.

Regarding ordinances, in general, town and city councils are
required to notify farmers concerning local actions that have a
direct and significant impact on agricultural operations.[413] Also,
in the early 1990s,[414] the Rhode Island legislature insulated agricultural operations from certain *specific* zoning ordinances that
regulate places for keeping animals including manure removal
and driving animals through the highways.[415] Moreover, farmers may place a seasonal direction sign or display on the state's
right of way, provided it conforms with local zoning ordinances
and is promptly removed at the end of the season.[416]

All is not doom and gloom in Rhode Island. The Supreme
Court of Rhode Island broadly interpreted the prohibition
against restrictive ordinances when it ruled against a township
that sought to use an earth removal ordinance and a zoning ordinance to stop the excavation of an irrigation pond on a
seventy-acre turf farm.[417] The town argued, among other things,
that the RTFA only applied to nuisance and not to earth removal
ordinances.

---

[412] This is the reading given by Attorney Alexadra Rawson on her website, available
at alexandrarawson.providence.wikispaces.net/What%27s+Protected%3F (last visited 22 Oct. 2016); however, I am not entirely convinced how the language including
agritourism specifically fits into the statutory scheme given the severe limitations in
the list of activities protected from nuisance actions, and later efforts to revise the
law which appear to recognize the same concerns. See 2015 R.I. H6100, available at
webserver.rilin.state.ri.us/billtext15/housetext15/h6100.pdf.

[413] R.I. Gen Laws § 2-23.1-1(e)(2016).

[414] 1990 R.I. ALS 145.

[415] R.I. Gen. Laws § 23-19.2-1 (2016).

[416] *Id.*

[417] *Town of N. Kingstown v. Albert*, 767 A.2d 659 (R.I. 2001).

However, the Court noted that the case arose from a series of complaints from neighbors about excessive dust. The Court said, "[A]lthough we agree with the town that the Farm Act is applicable to nuisance actions, we are cognizant that the statute is a statement of policy by the Legislature that farming activities and activities incidental to the right to farm ought not to be arbitrarily prohibited on the ground that the activity is objectionable on the ground of nuisance to either surrounding landowners or the municipality where the farm is located." Because the town's removal and zoning ordinance was in direct conflict with the landowner's right to continue to farm that land and would make it impossible to supply water to their crops, the Court held against the township.

South Carolina: South Carolina Code Ann. §§ 46-45-10 to §46-45-80[418]

Existing farms in South Carolina are protected under the RTFA from nuisance complaints when non-agricultural land use expands into traditionally agricultural areas.

Under the RTFA, established agricultural facilities and agricultural operations may not become public or private nuisances by any changed conditions in or about the locality of the facility or operation. However, the RTFA doesn't apply to negligent, improper, or illegal operation of an agricultural facility or operation.[419] Nor does it apply to polluting stream waters or to flooding.[420]

Agricultural operations are "established" the day they commence. There is not a one or two year requirement. Moreover, subsequent expansions or the adoption of new technology won't nix your "established" date.[421]

---

[418] www.scstatehouse.gov/code/t46c045.php.
[419] S.C. Code Ann. § 46-45-70 (2016).
[420] S.C. Code Ann. § 46-45-50 (2016).
[421] S.C. Code Ann. § 46-45-40 (2016).

There is a "commercial" requirement for agricultural facilities and for livestock operations to be protected under the RTFA. Agricultural *facilities* are defined as those used for *commercial* production or processing and agricultural operations. On the other hand, there is an extensive list of protected "agricultural *operations*." There is only a commercial requirement for livestock, and not for other operations like growing and harvesting crops or raising bees.

Moreover, "agricultural operations" includes "the application of existing, changed, or new technology, practices, processes, or procedures to an agricultural operation" as well as the operation of roadside markets.

Additionally, new local ordinances are void if not identical to state law regarding agricultural operations and facilities or if they seek to make an agricultural facility or operation a nuisance[422] (except for swine CAFOs which are handled in a totally separate statutory scheme[423]). However, the preemption of local ordinances does not apply if the nuisance results from negligent, illegal, or improper operation. Nor does the preemption apply to agricultural facilities or operations located within the corporate limits of a city.[424]

It bears repeating, not every ordinance relating to agricultural use is void, only those contrary to state law or that would make the facility or operation a nuisance.

For instance, certain setback and siting requirements are still okay.[425] And to the extent it complies with state law, counties aren't precluded from determining whether an agricultural use

---

[422] S.C. Code Ann. § 46-45-60 (2016).
[423] Confinement swine feeding operations are regulated separately in the code. S.C. Code Ann. § 47-20-165 et seq. With regard to swine CAFOs, all local ordinance that are not identical to state provisions are precluded. S.C. Code Ann. § 46-45-10 (2016).
[424] *Id.*
[425] S.C. Code Ann. § 46-45-80 (2016).

is a permitted use under the county's land use and zoning authority.[426]

South Dakota: South Dakota Codified Laws § 21-10-25.2[427]

The RTFA, adopted in 1991, limits the circumstances under which *commercial* agricultural operations may be deemed a nuisance especially where nonagricultural land uses encroach.[428]

To that end, agricultural operations cannot be public or private nuisances based on any changed condition in the locality of the operation after the facility has been in operation for more than one year, provided it wasn't a nuisance at the time the operation began.[429]

The term "agricultural operations" has a rather limited definition in South Dakota as compared to the extensive lists in other states. Only the following are protected: "any facility used in the production or processing for *commercial* purposes of crops, timber, livestock, swine, poultry, livestock products, swine products or poultry products."[430]

Reasonable expansions to agricultural operations are protected under the RTFA in terms of acreage and animals, provided all county, municipal, state, and federal environmental codes are met. Moreover, RTFA protection isn't lost by sale, gift, or inheritance of the operation; nor is it lost by a temporary cessation of farming or a diminution in size of the operation.[431]

As to zoning ordinances, township regulations cannot be in excess of or contrary to the county zoning commission; rather,

---

[426] S.C. Code Ann. § 46-45-60 (2016).
[427] www.sdlegislature.gov/Statutes/Codified_Laws/DisplayStatute.aspx?Type=Statute&Statute=21-10-25.2.
[428] S.D. Codified Laws § 21-10-25.1 (2016).
[429] S.D. Codified Laws § 21-10-25.2 (2016).
[430] S.D. Codified Laws § 21-10-25.3 (2016).
[431] S.D. Codified Laws § 21-10-25.2 (2016).

state and county regulations must be uniformly applied.[432] Hence a commercial hog feedlot couldn't be prevented by township regulations where it was allowed by the county zoning commission.[433] The court noted that the statute does not expressly require that expansions comply with "township" codes, laws, or regulations (only with county, municipal, state, and federal).[434]

Many county ordinances incorporate Right to Farm and require notice to residential properties in agricultural districts, that residents should be prepared to put up with the following:

> [The] cultivation, harvesting, and storage of crops; livestock production; ground rig or aerial application of pesticides or herbicides; the application of fertilizer, including animal waste; the operation of machinery; the application of irrigation water; and other accepted and customary agricultural activities conducted in accordance with Federal, State, and County laws. Discomforts and inconveniences may include, but are not limited to: noise, odors, fumes, dust, smoke, burning, vibrations, insects, rodents, and/or the operation of machinery (including aircraft) during any twenty-four hour period.[435]

It is not clear where the county came by the language of "other accepted and customary agricultural activities," as such is not included in the RTFA itself.

Regardless, the RTFA won't protect against negligent or improper operation.[436] Nor does it protect against pollution of streams or overflow of land.[437] Neither does it apply to any farm located within city limits prior to January 1, 1991.[438]

---

[432] *Welsh v. Centerville Tp.*, 595 N.W.2d 622, 629 (S.D. 1999).
[433] *Id.*
[434] *Id.* at n. 10.
[435] 2014 Revised Zoning Ordinance for Tucker County, South Dakota, available at turner.sdcounties.org/files/2010/09/Final-2014-Revised-Zoning-Ordinance_Effective-10-29-141.doc.
[436] *Id.*
[437] S.D. Codified Laws § 21-10-25.4 (2016).
[438] S.D. Codified Laws § 21-10-25.5 (2016).

Finally, the RTFA says that a farming defendant *shall* recover attorney fees and costs for frivolous lawsuits.[439]

Tennessee: Tennessee Code Annotated §§ 43-26-103 & 44-18-102

Farm operations in Tennessee get a rebuttable presumption that they aren't a public or private nuisance, which can only be overcome if the plaintiff proves by preponderant evidence either (1) by *expert testimony* that the farm doesn't conform to generally accepted agricultural practices; or (2) that the operation doesn't comply with applicable statutes or rules, especially those issued by the department of agriculture or the department of environment and conservations.[440]

Hemp production is treated a bit differently and requires licensing and inspection.[441]

As with several states, Tennessee breaks CAFOs out of the RTFA, handling them separately. Specifically, feedlots, dairy farms, or poultry productions get an absolute defense against persons whose ownership or actual use of the land occurred after the feedlot, dairy farm, or poultry production house began operating, provided that operation is in compliance with all applicable state regulations and all applicable zoning ordinances.[442]

Moreover, the statute goes on to specifically state that normal noises, odors, and appearances, etcetera, of feedlots, dairy farms, and poultry production houses are not actionable in nuisance if a person moved in subsequent to the date of operation.[443]

Going back to the RTFA for a moment, it used to be that "new type of farming operations" (meaning expansions, additions, or

---

[439] S.D. Codified Laws § 21-10-25.6 (2016).
[440] Tenn. Code Ann. § 43-26-103 (2016).
[441] Tenn. Code Ann. § 43-26-103(b) and (c) (2016).
[442] Tenn. Code Ann. § 44-18-102 (2016).
[443] Tenn. Code Ann. § 44-18-102 (2016).

operations that were materially different after the plaintiff acquired his land) were not protected by the RTFA till after they'd been in operation one year. However, that section was deleted by a 2016 amendment.[444] Therefore, one might argue that expansions, changes in the type of operation, and the adoption of new technologies are now protected under the RTFA.

The definition of "farm operations" under the RTFA is quite broad and includes selling produce at roadside stands or farm markets; marketing in conjunction with the production of farm products; as well as other definitions of "agriculture."[445]

Agritourism is included in the definition of "agriculture."[446] And in 2014, the code was amended[447] to include "[e]ntertainment activities conducted in conjunction with, but secondary to, commercial production of farm products and nursery stock, when such activities occur on land used for the commercial production of farm products and nursery stock."[448] This amendment took place after the Tennessee Supreme Court overruled an appellate court who had decided that operation of a pumpkin patch, a corn maze, and music concerts were protected activities.[449] The appellate court held the farm was protected from the application of the local zoning laws by the RTFA because their farm activities were sufficient to meet the definition of agritourism; however, the supreme court disagreed, saying the concerts bore no relation to the production of cattle, corn, vegetables, strawberries, or pumpkins at the farm.[450] The legislature appears to have enshrined the appellate court decision protecting agritourism.

---

[444] 2016 Tenn. ALS 728.
[445] Tenn. Code Ann. § 43-26-102 (2016).
[446] Tenn. Code Ann. §§ 1-3-105(2)(A)(i-iii) & 43-1-113(b)(1).
[447] 2014 Tenn. ALS 581.
[448] Tenn. Code Ann. §§ 1-3-105(2)(A)(i-iii) & 43-1-113(b)(1).
[449] *Shore v. Maple Lane Farms, LLC*, 2012 Tenn. App. LEXIS 229 (Tenn. Ct. App. Apr. 11, 2012), *rev'd*, 411 S.W.3d 405 (Tenn. 2013).
[450] *Id.*

In addition, the legislature explicitly states that the RTFA is to be broadly construed.[451]

As to zoning ordinances, there are certain exemptions from zoning ordinances for agriculture. First, local government zoning and police power regulations are circumscribed by and may not conflict with state laws.[452]

Moreover, buildings or structures incident to agriculture and on land devoted to agricultural use are generally exempt from zoning requirements except where they abut highways, airports, or parks.[453] Likewise, a county's delegated authority to exercise general police powers does not extend "the power to prohibit or regulate normal agricultural activities."[454]

These various statutes all add up to broad protection for the Right to Farm under Tennessee law.

Texas: Texas Agriculture Code § 251.004[455]

In Texas, no nuisance actions may be brought against agricultural operations that have lawfully been in operation for a year or more prior to the date the action is brought, provided that the conditions or circumstances complained of have existed substantially unchanged since the established date of operation.[456] Moreover, subsequent expansion of a physical facility counts as a separate, independent date of operation.[457]

According to the courts in Texas, agricultural operations don't have to prove when they actually started; rather, they only need to show that the circumstances or condition complained of

[451] Tenn. Code Ann. § 43-26-104 (2016).
[452] *421 Corp. v. Metropolitan Gov't of Nashville & Davidson County*, 36 S.W.3d 469, 475 (Tenn. Ct. App. 2000).
[453] Tenn. Code Ann. § 13-7-114 (2016).
[454] Tenn. Code Ann. § 5-1-122 (2016).
[455] www.statutes.legis.state.tx.us/Docs/AG/htm/AG.251.htm.
[456] Texas Agric. Code § 251.004 (2016).
[457] Texas Agric. Code § 251.006 (2016).

had remained substantially unchanged more than one year before suit was filed in order to avoid nuisance *and trespass*
claims.[458]

Agricultural operations include: cultivating the soil; producing crops for human food, animal feed, planting seed, or fiber;
floriculture; viticulture; horticulture; silviculture; wildlife management; raising or keeping livestock or poultry; and planting
cover crops or leaving land idle for participation in government
programs or as part of crop or livestock rotation;[459] and, by court
opinion, grain storage facilities.[460]

In addition, construction or maintenance of an improvement
on agricultural land (such as a pen, barn, or fence) won't subject
a farmer to state, governmental, or private liability, nor will it be
a nuisance, provided the improvement didn't violate state or local rules when it was built and provided it doesn't obstruct the
flow of water, light, or air to other land.[461]

As regards the one-year restriction on bringing a lawsuit, according to the Texas Supreme Court, "it does not matter when
the complaining party discovers the conditions or circumstances
constituting the basis for the nuisance action. Instead, the rele-
vant inquiry is whether the conditions or circumstances
constituting the basis for the nuisance action have existed for
more than a year."[462]

In that case, a sheep farm expanded its physical facilities,
adding a 5,800 sheep CAFO within 135 feet of plaintiff's house.
The plaintiffs claimed they didn't discover the nuisance till the
heat of summer brought odors and flies. But the Supreme Court

---

[458] *Ehler v. LVDVD, L.C.*, 319 S.W.3d 817 (Tex. App. 2010). *See also, Aguilar v. Trujillo*, 162 S.W.3d 839 (Tex. App. 2005); *Bergin v. Tex. Beef Group*, 339 S.W.3d 312 (Tex. App. 2011).
[459] Texas Agric. Code § 251.002 (2016).
[460] *Cal-Co Grain Co. v. Whatley*, No. 13-05-120-CV, 2006 Tex. App. LEXIS 7536 (Tex. App. Aug. 24, 2006).
[461] Texas Agric. Code § 251.003 (2016).
[462] *Holubec v. Brandenberger*, 111 S.W.3d 32 (Tex. 2003).

said the date of operation is the date the conditions or circumstances existed, not the date they're discovered. Hence, the plaintiffs had to have brought the case within a year of the date the CAFO began operating, which they hadn't.

It may seem like an odd ruling, because, how could you know the operation was a nuisance before it becomes a nuisance? However, the decision relies on a legal technicality[463] based on a precise reading of the statute. I won't go into the details. Suffice it to say, if you don't get sued for a year, in Texas, you should be all right.

If it makes you feel any better, on remand, the jury found against the CAFO, finding that it had been a nuisance since July 1997; and constituted a trespass; and that the landowners had been negligent in the location, construction, or operation of their feedlot; and that the landowner had fraudulently concealed the fact that they were going to operate a sheep feedlot; and that the harm to plaintiffs had resulted from malice or fraud. Wow![464]

Money damages including expensive punitive damages were awarded against the landowners. On appeal, the court reversed a permanent injunction, where, in fairness to the farmers, the farm had been used to raise a limited number of sheep prior to the CAFO and even the CAFO itself might conceivably be relocated to another portion of defendant's 450 acre farm without harm to the complainants.[465]

Moving back to the RTFA, it doesn't prevent the state from

---

[463] The statute is a statute of repose instead of a statute of limitations. One difference is, a statute of repose is independent of the claim's accrual or discovery. *Trinity River Auth. v. URS Consultants*, 889 S.W.2d 259, 261 (Tex. 1994); thus, "statutes of repose not only cut off rights of action within a specified time after they accrue, but also they may even cut off rights of action before they accrue at all." *Holubec v. Brandenberger*, 111 S.W.3d 32, 37 (Tex. 2003), *citing Johnson v. City of Fort Worth*, 774 S.W.2d 653, 654 n.1 (Tex. 1989) (per curiam). Put another way, statutes of repose "fix an outer limit beyond which no action can be maintained"—period, end of story. *Holubec v. Brandenberger*, 111 S.W.3d 32, 37 (Tex. 2003)

[464] *Holubec v. Brandenberger*, 214 S.W.3d 650, 654 (Tex. App. Austin 2006).

[465] *Id.* at 659.

protecting the public's health, safety, and welfare, nor does it re-move a municipality's authority to enforce state law.

The RTFA doesn't protect a farm where it violates applicable federal, state, or local statutes or governmental requirements. It doesn't cover raising fighting cocks, sorry.[466] Nor does it cover the use of a propane cannon where that use had *not* remained substantially unchanged since the operation began.[467]

That ruling highlights the need to keep detailed records as to equipment purchase and usage, and it highlights the fact that modifications to farm practices may eliminate protection under the RTFA.[468]

As for attorney's fees and costs, the Texas statute is quite gen-erous. A person who sues an agricultural operation *is liable* to the agricultural operation for attorney's fees and costs. However, the RTFA doesn't authorize a farm to recover for costs and ex-penses incurred while defending an action brought by the Texas Department of Health or a local prosecuting attorney.[469]

As with every jurisdiction except Iowa, the Texas appellate court held that the RTFA is not an unconstitutional taking with-out compensation under the Fifth Amendment.[470]

With regard to zoning, in addition to the exception for im-provements on agricultural land mentioned above, according to the RTFA, cities are also limited from imposing zoning or other restrictions on agricultural operations outside their corporate boundaries or on farms later annexed, with the exception of cer-tain public health or other concerns including the discharge of

---

[466] *Hendrickson v. Swyers*, 9 S.W.3d 298 (Tex. App. 1999).

[467] *Reeves v. Hooton*, 2013 Tex. App. LEXIS 11012 (Tex. App. Aug. 29, 2013).

[468] Tiffany Dowell, "Texas Right To Farm Act Does Not Protect Farmer Using Pro-pane Cannon," Texas Agricultural Law Blog, Sept. 2, 2013, agrilife.org/texasaglaw/2013/09/02/texas-right-to-farm-act-does-not-protect-farmer-using-pro-pane-cannon.

[469] Tex. Atty. Gen. Op. MW-544, 1982 WL 173915 (Tex. A.G. 1982), available at www.texasattorneygeneral.gov/opinions/opin-ions/46white/op/1982/pdf/mw0544.pdf.

[470] *Barrera v. Hondo Creek Cattle Co.*, 132 S.W.3d 544 (Tex. App. 2004).

firearms.[471] There may be some of the other exceptions for zoning for buildings on agricultural land and the like, but those
weren't readily accessible and at any rate are beyond the scope
of this survey.

Utah: Utah Code Annotated § 17-41-403[472] & § 76-10-803[473]

In agricultural protection areas in Utah, normal agricultural uses
and activities are afforded the highest priority use status and are
protected by a statutory scheme that includes an RTFA.

Hence, localities are instructed that in agriculture protection
areas, laws and ordinances defining public nuisances should ex-
clude any activity or operation that is using sound agricultural
practices unless there's a direct impact on public health or
safety.[474]

In fact, for agricultural protection areas, local governments
should pass no unreasonably restrictive local laws, ordinances,
or regulations concerning farm structures or practices absent
public health or safety concerns.

There is also a limitation on their ability to change the zoning
designation within an agricultural protection area without written consent from all the landowners within the area affected by
the change, (making it difficult, one would assume, to sell off a
piece of farmland to make a residential development next to
farming neighbors.) This should prevent residents from coming
to the nuisance.

It's not only local governments that are limited. State agencies are similarly restricted as to passing new rules or modifying
existing rules.[475]

---

[471] Tex. Agric. Code §§ 251.002 & 251.005 (2016).
[472] www.le.utah.gov/xcode/Title17/Chapter41/17-41-S403.html.
[473] www.le.utah.gov/xcode/Title76/Chapter10/76-10-S803.html.
[474] Utah Code Ann. § 17-41-402(1) (2016).
[475] Utah Code Ann. § 17-41-404 (2016).

Moreover, as to civil actions for nuisance or a criminal action for public nuisance, agricultural activities are provided a complete defense if they're conducted within an agricultural protection area and aren't in violation of any applicable federal, state, or local law or regulation *or* were conducted according to sound agricultural practices.[476]

Farmers who want to have their land designated as an agricultural protection area are encouraged to file with the county council and follow the subsequent process.

As to a criminal action for public nuisance, activities conducted in the normal and ordinary course of "agricultural operation" and conducted in accordance with sound agricultural practices are presumed to be reasonable and not constitute a public nuisance. In addition, agricultural operations undertaken in conformity with federal, state, and local laws and regulations, including zoning ordinances, are presumed to be operating within sound agricultural practices.[477]

Under the civil code concerning nuisance, "agricultural operation" is defined as any activity engaged in the *commercial* production of crops, orchards, aquaculture, livestock, poultry, livestock products, poultry products, and the facilities, equipment, and property used to facilitate the activity.[478]

Let me be a bit precise and perhaps boring for a moment. Those are three separate code sections: the RTFA, the law on public nuisance, and the law on private nuisance. The interaction between these sections isn't entirely clear. For instance, the term "agricultural operation" has its definition in an unrelated, separate subsection of the civil nuisance code. One might assume that definition is meant to be ported into the RTFA. However, the RTFA itself does not contain the term "agricultural operation." The RTFA only refers to "agricultural activities."

---

[476] Utah Code Ann. § 17-41-403(2) (2016).
[477] Utah Code Ann. § 76-10-803 (2016).
[478] Utah Code Ann. § 78B-6-1101 (2016).

Therefore, one might argue that a farm doesn't need to be *commercial* (a requirement of the agricultural operation definition) to qualify for RTFA protection. Instead, the term "agricultural operation" shows up in the criminal, public nuisance code. However, it is typically assumed by courts that terms *do not* have the same meaning when used in such different portions of the code. For example, a definition of "consent" for the purposes of a civil actions would *not* apply to the term "consent" under the criminal code.

Regardless, getting back to the RTFA itself, developers are required to give notice on any plat within 300 feet of an agricultural protection area, that future landowners may be subject to normal agricultural uses and activities.[479]

Eminent domain and state development projects are substantially restricted as concerns land in an agriculture protection area that is being used for agricultural production.[480]

Finally, as with several states, Utah has a law criminalizing undercover recording operations. It's a misdemeanor to record an image or sound on private property without an owner's consent by leaving a device behind; or to obtain access under false pretenses or by trespass; or to insinuate yourself into an operation as an employee to record an image or sound knowing the owner doesn't allow it.[481] That law is being challenged in federal court as being unconstitutional.[482] A federal court in Idaho found

---

[479] Utah Code Ann. § 17-41-403 (2016).
[480] Utah Code Ann. § 17-41-405 (2016).
[481] Utah Code Ann. § 76-6-112 (2016).
[482] Willy Blackmore, "Is Utah's Ag-Gag Law Unconstitutional?" Take Part, July 24, 2013, www.takepart.com/article/2013/07/24/utahs-ag-gag-law-unconstitutional.

a similar law unconstitutional,[483] while a court in Wyoming totally disagreed[484]. Another similar law is being challenged in North Carolina. We'll have to wait and see what happens in Utah, but the Wyoming court seems to have the better of the legal argument.

That statute also creates a private civil cause of action not only against the perpetrators but potentially against those who worked with them including their support and distribution networks.

Regardless, it is important, then, for employers to make their policies about image and sound recording and as well as data security clear to their employees. Farms should make their policy on recordings or lack thereof clear to their employees, document the process, and be sure to enforce it consistently. For instance, if you let your employees post selfies to social media, you may have a hard time suing an "employee" who took it upon themselves to send undercover photos to an overzealous animal rights group.

Vermont: Vermont Statutes Annotated title 12, section 5753[485]

The RTFA in Vermont states that agricultural activity conducted in a way that is consistent with good agricultural practices and established prior to surrounding nonagricultural activities, shall be entitled to a rebuttable presumption that the activity does not constitute a nuisance, provided it is conducted in conformity with federal, state, and local laws and regulations (including "required agricultural practices") and provided that it hasn't

---

[483] *Animal Legal Def. Fund v. Otter*, 118 F. Supp. 3d 1195 (D. Idaho 2015).
[484] *Western Watersheds Project v. Michael*, 2016 U.S. Dist. LEXIS 88843 (D. Wyo. July 6, 2016) Criticizing the Idaho court's decision, the court in Wyoming said, "There is no precedent to support the premise that whistleblowers somehow have an elevated First Amendment right to make audiovisual recordings on private property without permission. No matter how virtuous or important one may view a whistleblower's motives or actions, the ends do not justify the means of trespass." *Id.* at *28.
[485] legislature.vermont.gov/statutes/chapter/12/195.

significantly changed since the commencement of the prior surrounding nonagricultural activity.

This presumption may be rebutted by a showing that the activity has a substantial adverse effect on the public health and
safety or has a noxious and significant interference with the use
and enjoyment of the neighboring property.[486]

As to what constitutes good agricultural practices, The Sec-
retary of Agriculture, Food, and Markets is authorized to define
"required agricultural practices" after conducting a public hearing.[487]

The RTFA reflects changes[488] made following a key decision
by the Vermont Supreme Court. In that case, plaintiffs bought a
200-year-old homestead on an apple orchard that later significantly expanded, causing complaints of excessive noise, light
glare, and fumes.[489] The lower court held against the homeowners, saying that they'd moved to the nuisance. Moreover, both
the agriculture commissioner and the zoning administrator said
that the orchard's activities conformed to state and local requirements and were reasonable agricultural activities. Finally, the
lower court found that plaintiffs had not raised an issue of public
safety or health.

However, the Supreme Court of Vermont held that the RTFA
did *not* bar the homeowner's suit because the agricultural activities that gave rise to the plaintiffs' nuisance began *after* the
plaintiff's had already purchased the home. The supreme court
also said, even if the orchard complied with state rules, that
didn't necessarily mean it wasn't creating a nuisance.

Following the supreme court case, the substantive portion of
the RTFA was amended as well as the legislative purposes and

[486] Vermont Stat. Ann. tit. 12, § 5753 (2016).
[487] Vermont Stat. Ann. tit. 10 § 1259(f) (2016); agriculture.vermont.gov/water-qual-
ity/regulations/rap/archive.
[488] 2003 Vt. ALS 149.
[489] *Trickett v. Ochs*, 838 A.2d 66 (Vt. 2003).

findings section. That section was amended to protect "the initi-
ation of new agricultural activities" and to recognize that "farms
will likely change, adopt new technologies, and diversify into
new products, which for some farms will mean increasing in
size."[490]

The Vermont RTFA defines protected "agricultural activities"
in the usual ways but also adds to that list the on-site production
and sale of agricultural products as well as the production of fuel
or power from agriculture or waste produced on site.[491]

The RTFA doesn't limit the authority of the state or local
boards to abate nuisances affecting the public health.

As might be expected, the RTFA won't protect you if your
hog feeding operation was established to retaliate against and
harass your neighbors, sorry.[492]

Moreover, Vermont joins a long list of states that do not allow
plaintiffs to avoid the RTFA by characterizing an action as trespass rather than nuisance.[493] In a case on point, the court held
that drifting pesticides–as particles on the ambient air—don't
have the physical impact necessary to be a "trespass."[494]

In terms of zoning ordinances, towns can establish agricultural districts and prohibit other types of development.[495] In
addition, "required agricultural practices" (RAPs) including the
construction of farm structures are exempt from local zoning
and development bylaws, though structures still have to comply
with setback requirements.[496] Farming is also exempt from the
Act 250 permitting process.[497] On this score, it *seems* as though

[490] Vermont Stat. Ann. tit. 12, § 5751 (2016).
[491] Vermont Stat. Ann. tit. 12, § 5752 (2016).
[492] *Coty v. Ramsey Associates, Inc.*, 546 A.2d 196 (1988).
[493] *John Larkin, Inc. v. Marceau*, 184 Vt. 207 (2008).
[494] *Id.*
[495] Vt. Stat. Ann. tit. 24 § 4414 (2016).
[496] Vt. Stat. Ann. tit. 24 § 4413 (2016).
[497] Vt. Stat. Ann. tit. 10 § 6001 (2016).

Vermont farmers have lost some of the broad protection afforded to them in the past.[498] For instance, the statute exempts
only RAPs and structures; meanwhile, the RAPs are pretty ane-
mic and limited solely to water quality.[499]

Virginia: Virginia Code Annotated § 3.2-302[500]

Bona fide agricultural operations are protected from private and
public nuisance actions, provided such operations are conducted in accordance with existing best management practices
and comply with existing laws and regulations of the Commonwealth. However, the RTFA doesn't protect negligent or
improper operations, nor against suits involving pollution of the
waters of any stream or any overflow of lands. Nor does it apply
to negligence or trespass actions.[501]

The RTFA further provides that any and all ordinances of
any unit of local government, present or future, that would make
the operation of any such agricultural operation a nuisance or
providing for abatement shall be null and void, unless the nuisance results from negligent or improper operation.

"Agricultural operation" means any operation devoted to the
bona fide production of crops, or animals, or fowl including the
production of fruits and vegetables of all kinds; meat, dairy, and
poultry products; nuts, tobacco, nursery,[502] and floral products;
and the production and harvest of products from silviculture activity.[503] However, according to an opinion of the attorney

---

[498] *See* Garrett Chrostek, "A Critique Of Vermont's Right-To-Farm Law And Pro-
posals For Better Protecting The State's Agricultural Future" 36 Vt. L. Rev. 233, 257
(Fall 2011); and see how things used to be in the court's discussion in *Trickett v. Ochs*,
838 A.2d 66, 69-70 (Vt. 2003).
[499] agriculture.vermont.gov/water-quality/regulations/rap/archive.
[500] law.lis.virginia.gov/vacode/title3.2/chapter3/section3.2-302.
[501] *Wyatt v. Sussex Surry, LLC*, 482 F. Supp. 2d 740 (E.D. Va. 2007).
[502] *Layng v. Gwinn*, 52 Va. Cir. 71 (2000) (holding that a plant nursery qualified as an
"agricultural use" under an earlier version of the Virginia Right to Farm Act).
[503] Va. Code Ann. § 3.2-300 (2016).

general, the RTFA doesn't protect aquaculture.[504] The AG's opinion is that "animals" in the list "crops, or animals, or fowl" doesn't apply to things like oysters. The opinion isn't binding on the judicial branch (though it may be followed by agencies under the executive branch). A court might find the AG's opinion persuasive.

FTCLDF member Anthony Bavuso has just such a case pending before the state supreme court challenging local ordinances that are affecting his oyster farm.[505]

Indeed, starting in 1995, the Virginia RTFA exempts production agriculture activity in agricultural districts or classifications from any special exception or special use permit.[506] However, localities may adopt setback requirements, minimum area requirements, and other requirements. [507] (In a list like that, "other requirements" is restricted to things like setback and minimum area requirements.) A zoning ordinance may not unreasonably restrict or regulate farm structures or farming in an agricultural district or classification unless such restrictions are related to public health, safety, and general welfare.[508] Hence, the RTFA exemptions apply to "areas currently zoned as agricultural districts or classifications, as well any other areas in which the zoning provisions allow for agricultural activity."[509]

So, for example, a log yard was protected by the RTFA from enforcement of a zoning ordinance. Operation of the log yard in a "Residential Agricultural" zoning district. It's operation wasn't a zoning violation, as such use involved silviculture, which was included in definition of agriculture, and agriculture

---

[504] 2012 Va. AG Lexis 11.
[505] *County of York v. Bavuso*, 2016 Va. LEXIS 48 (Va. Apr. 12, 2016).
[506] Va. Code Ann. § 3.2-301 (2016); Va. Code Ann. § 15.2-2288 (2016).
[507] *Id.*
[508] *Id.*
[509] 2011 Va. AG Lexis 38.

was permitted by right[510] in that district.[511] The district tried to shut the log yard down, claiming it was a "distribution center" rather than an agricultural use, but the court disagreed, citing the plain language of the RTFA.

In contrast, in 2013, the Virginia Attorney General issued an opinion stating that, as to a "Rural *Residential*" zoning district, the authorization of agricultural uses even "by right" *does not* make it an "agricultural district" under the RTFA so as to exempt production agriculture activities from zoning ordinances.[512] Rather, "a Rural *Residential* zoning district would be considered a residential classification" says the AG. (Remember, activity in agricultural "districts or *classifications*" is exempt. One might argue allowing agriculture "by right" in a Rural Residential *district* would change the *classification*. And indeed there is some case law in other states that makes the distinction between "district" and "classification."

The AG nevertheless says, Rural Residential zoning is a classification that "promotes residential uses in rural areas." [513] Hence, a municipality may regulate however it sees fit including restricting or allowing all agricultural activities.[514]

In such cases, one must then pay very careful attention to the zoning classifications and look for any other areas in which the zoning provisions allow for agricultural activity to see whether the RTFA applies. That's especially true since the RTFA limits municipal rules, as an opinion by the Attorney General explains:

[510] The land was designated by the county as an "A-3 zoning district" used "to provide for the continued practice of agriculture, farm operations, agriculturally related and home based businesses, low density residential developments, preferably in a hamlet subdivision pattern, and other uses in a predominantly rural environment. The district also permits direct marketing of farm products and services." *Buckley v. Loudoun County Bd. of Zoning Appeals*, 59 Va. Cir. 150 (2002).
[511] *Buckley v. Loudoun County Bd. of Zoning Appeals*, 59 Va. Cir. 150 (2002).
[512] 2013 Va. AG Lexis 41.
[513] *Id.*
[514] *Id.*

The powers of county boards of supervisors in the Common-wealth are limited to those "conferred expressly or by necessary implication." "'This [principle] is a corollary to Dillon's Rule that municipal corporations have only those powers expressly granted, those necessarily or fairly implied therefrom, and those that are essential and indispensable.' "Any doubt as to the existence of a power must be resolved against the locality. Accordingly, because local governments are creatures of the Commonwealth and thus subordinate, they possess only those powers that the General Assembly has conferred upon them. Thus, where express statutory provisions prohibit localities from adopting ordinances regulating certain matters, there is legislative intent to remove from local governments the authority to do so.

An ordinance is inconsistent with state law if state law preempts any local regulation in the area. The General Assembly, in its enactment of the Right to Farm Act, expressly limited the circumstances under which an agricultural operation is deemed a nuisance. Additionally, it prohibited a county from adopting an ordinance requiring a special exception or special use permit for agricultural activity (with the exception of certain setback, minimum area and other requirements applicable to the land on which the agricultural activity is occurring). . . . Although the Right to Farm Act does not specifically prohibit all local regulation of industrial farming, any restrictions must "bear a relationship to the health, safety and general welfare" of the locality's citizens. Therefore, unless it is attributable to the safeguarding of a locality's inhabitants, a local ordinance regulating industrial farming is preempted by the Right to Farm Act.

Based on the above, it is my opinion that a county does not have the authority to adopt an ordinance, such as the one proposed, that limits the circumstances under which agricultural operations may be deemed to constitute a nuisance, trespass, or other interference with the reasonable use and enjoyment of land.[515]

Washington: Revised Code Washington § 7.48.305[516]

Agricultural activities conducted on farmland,[517] if consistent

---

[515] 1998 Op. Atty Gen. Va. 13.
[516] app.leg.wa.gov/rcw/default.aspx?cite=7.48.305.
[517] The RTFA includes "forest practices" as well.

with good agricultural practices and established prior to surrounding nonagricultural activities, are presumed to be reasonable and shall not be found to constitute a nuisance unless the activity or practice has a substantial adverse effect on public health and safety. Moreover, agricultural activities that comply with all applicable laws and rules are *presumed* to have no adverse effect on public health and safety. In addition, agricultural activities on farmland may not be restricted to certain hours of the day or days of the week. That being said, nothing in the RTFA prevents other sorts of actions from being filed against farms.[518]

The RTFA also extends special protection to land that is sustaining a crops of trees even though it appears inactive. This after a court held mere ownership of forest land was insufficient to constitute a preexisting "practice" entitling the landowner to nuisance immunity. The landowner's ownership of the forest land predated the community's location, but the logging did not; hence the operation did not receive RTFA protection when the logging activities caused a nuisance.[519] After that ruling, the RTFA was amended as to forest activities.

Recall, the RTFA applies where a farm practice was occurring *before* a change in nonagricultural use. Hence the court didn't protect the farmer who engaged in deforestation that caused avalanches affecting a previously established residential community.[520] Other courts held similarly where a residence existed before a rock quarry[521]; and where a farmer used propane

---

[518] Rev. Code Wash. (ARCW) § 7.48.305 (2016).
[519] *Alpental Cmty. Club v. Gym. Soc'y*, 111 P.3d 257 (Wash. 2005) (avalanches occurring after upland property owner engaged in deforestation activity).
[520] *Id.*
[521] *Gill v. LDI*, 19 F. Supp. 2d 1188 (W.D. Wash. 1998) (In addition, the quarry had not engaged in "good farming practices" since it had violated several water quality laws.).

cannons and cherry guns after the adjoining residential neighborhood was established.[522]

The Right to Farm doesn't apply to new farms that startup near nonagricultural or residential areas. Nor does it mean a "right" to grow whatever you like, wherever you like, such as raising chickens in your urban backyard. You'd have to look to other state and local rules to determine what's allowable on your land.

Moreover, in Washington, existing farms need to be especially careful about any changes made on the farm itself *after* residences move into the neighborhood, else they might lose protection under the RTFA. In fact, the state supreme court has said that new or expanded activities, after local development, are not exempt from a nuisance liability; rather the nuisance exemption only applies to lawsuits that arise because of encroaching urbanization.[523]

That being said, the term "agricultural activity" in addition to including the usual activities as well as bees, roadside stands, and farm markets, also includes "conversion from one agricultural activity to another, including a change in the type of plant-related farm product being produced" and the "use of new practices and equipment consistent with technological development within the agricultural industry."[524] Therefore, some changes might be protected under the RTFA.

For example, a mushroom farm that operated an indoor composting facility was protected by the RTFA because the compost activities cannot be separated from the mushroom growing; neither was it a "new or radically expanded activity" because, although the defendant's indoor composting facility was a new

---

[522] *Davis v. Taylor*, 132 P.3d 783 (Wash. Ct. App.), superseded, 132 P.3d 783 (Wash. Ct. App. 2006) (cherry guns and propane cannons).
[523] *Buchanan v. Simplot Feeders LTD. (In re Certification for Eastern Dist. of Wash.)*, 134 Wn.2d 673, 680-681 (Wash. 1998).
[524] Rev. Code Wash. (ARCW) § 7.48.310 (2016).

building, the composting of mushrooms was not a new agricultural activity nor was it an expanded activity since the defendant had always used compost in his operation.[525]

On the other hand, an appellate court upheld an injunction forbidding orchard owners from discharging excess irrigation water from their orchards past a certain point to prevent damage to adjacent property. The court explained that the legislative history of the RTFA indicated that the legislature did not intend the type of flooding that had occurred in this case to be given nuisance protection. The court said, even assuming "rill irrigation" was reasonable, flooding adjoining property wasn't what was intended by the nuisance exemption. The nuisance exception contemplated sounds, smells, dust, etc., which might interfere with use and enjoyment of life and property. The law didn't protect interference with possession of property, by actual destruction through escaping water, which is categorically different according to that court.[526]

As for local zoning ordinances and other rules, there was a tangential case where an appeals court in Washington upheld a decision by a board of commissioners denying a request to plat a residential subdivision next to an orchard based on health and safety concerns. The orchard sprayed pesticides, had noise day and night including twenty-foot anti-frost propellers as well as propane cannons; fires and smoke and noise and lights at night etcetera.[527] Moreover, the residents would be precluded by the RTFA from bringing a suit despite the existence of nuisance-like conditions. Given the effect on the general welfare of the subdivision's residents, the county had the authority to consider whether the proposed subdivision "was compatible with the lawful operation of the neighboring orchard."[528]

---

[525] *Vicwood Meridian P'ship v. Skagit Sand*, 98 P.3d 1277 (Wash. Ct. App. 2004).
[526] *City of Benton City v. Adrian*, 748 P.2d 679, 682 (Wash. Ct. App. 1988).
[527] *Kanna v. Benton County*, 95 Wash. App. 1011 (Div. 3 1999).
[528] *Id.*

We could find no language limiting application of zoning ordinances on agricultural activities. It seems like even non-residential farm structures require permits, unless otherwise allowed by the county.[529]

Finally, the RTFA says farmers may recover attorney fees and costs if they prevail and may even get exemplary damages if the court finds the case was initiated maliciously and without probable cause.[530] State agencies may likewise recover their costs if prompted to investigate by a complaint that was made maliciously and without probable cause.[531]

West Virginia: West Virginia Code § 19-19-4[532]

A proposed Right to Farm constitutional amendment nearly came up for a vote in 2016, having passed both houses. Perhaps it will be revived at a later date.

Regardless, the RTFA in West Virginia is quite broad. Agriculture conducted on land of five acres and bringing in $1,000 in annual profits is exempt from any complaint or right of action in any state court *nor* shall such conduct be deemed adverse to other use or uses of adjoining or neighboring land.

These broad protections are predicated on two conditions: (1) that the complainant not have owned adjoining or neighboring land before the agricultural operation complained of, and (2) the operation complained of hasn't caused or won't cause actual physical damage.[533]

Agriculture means the usual sort of activities including bees; however, it doesn't cover manufacturing, milling, or processing

---

[529] Pierce County allows unpermitted agricultural buildings of 600 square feet or less provided they follow some very specific guidelines.
www.piercecountycd.org/Blog.aspx?IID=29.
[530] Rev. Code Wash. (ARCW) § 7.48.315 (2016).
[531] Rev. Code Wash. (ARCW) § 7.48.320 (2016).
[532] www.legis.state.wv.us/wvcode/ChapterEntire.cfm?chap=19&art=19&section=4.
[533] W. Va. Code, § 19-19-4 (2016).

unless done by the producer. Agricultural land must be five or more acres and must produce products of $1,000 or more annually.[534]

A temporary cessation or change to a differing agricultural use shall not constitute abandonment or limit the change to any other agricultural use.[535]

The RTFA only applies to the right to conduct the practice of agriculture upon agricultural lands and doesn't excuse any other right or duty owed.[536]

There are some limited protections from zoning restrictions as to agricultural structures. Prior nonconforming uses are grandfathered in; moreover,

> [N]o zoning ordinance may prohibit alterations or additions to or replacement of buildings or structures owned by any farm . . . or the use of land presently owned by any farm but not used for agricultural purposes, or the use or acquisition of additional land which may be required for the protection, continuing development or expansion of any agricultural . . . operation of any present or future satellite agricultural . . . use.[537]

The West Virginia Farm Bureau, however, has identified some gaps in RTFA protection and recommends the adoption of an RTFA that will:

(1) Create a legal presumption that agriculture management practices are not a nuisance and are an expected part of the quiet enjoyment of property;
(2) Protect agriculture from nuisance lawsuits and or/complaints against generally accepted management practices;
(3) Exempt agriculture from noise ordinances, light ordinances, dilapidated building ordinances, and other nuisance ordinances;
(4) Establish a disclosure provision whereby the seller or his/her agent

---

[534] W. Va. Code, § 19-19-2 (2016).
[535] W. Va. Code, § 19-19-3 (2016).
[536] W. Va. Code, § 19-19-5 (2016).
[537] W. Va. Code § 8A-7-10 (2016).

and the county clerk are required to notify potential buyers of the property of this right to farm ordinance; and

(5) Exempt agriculture from zoning restrictions and other regulatory tools that may interfere with generally accepted agriculture management practices.[538]

Wisconsin: Wisconsin Statutes Annotated 823.08[539]

The RTFA was created in order, to the extent possible consistent with good public policy, to keep the law from hampering agricultural production or preventing the use of modern agricultural technology. The legislature also urged local units of government to use their zoning power to prevent conflict between agricultural and nonagricultural uses.

A commercial agricultural use or agricultural practice may not be found to be a nuisance provided the land had an agricultural use without substantial interruption before the plaintiff began the use of his property and the use or practice doesn't present a substantial threat to public health or safety. This applies regardless of whether a *change* in agricultural use or practice is alleged to have contributed to the nuisance.

The term "agricultural use" is defined as any of the following activities *conducted for the purpose of producing an income or livelihood*: 1. Crop or forage production; 2. Keeping livestock; 3. Beekeeping; 4. Nursery, sod, or Christmas tree production; 4m. Floriculture; 5. Aquaculture; 6. Fur farming; 7. Forest management; 8. Enrolling land in a federal agricultural commodity payment program or a federal or state agricultural land conservation payment program; or (b) Any other use that the department, by rule, identifies as an agricultural use.

Moreover, an "agricultural practice" is defined as "any activity associated with an agricultural use."[540]

---

[538] West Virginia Farm Bureau Policy Book 2016, www.wvfarm.org/download/PolicyBook.pdf.

[539] docs.legis.wisconsin.gov/statutes/statutes/823/08.

[540] Wisc. Stat. Ann. § 91.01(2) (2016).

Moreover, the RTFA limits damages for successful nuisance actions. Specifically, the relief granted may not "substantially restrict or regulate the agricultural use or agricultural practice" unless there is a *substantial* threat to public health or safety. Further, before ordering a defendant to take any actions to mitigate the effects of a nuisance, the court must request practicable suggestions from public agencies with expertise in agriculture and must give the defendant reasonable time (at least a year) to implement the court's directions unless there's a substantial threat to public health or safety. Finally, orders to mitigate may not mandate action that substantially and adversely affects the economic viability of the agricultural use unless there's a substantial threat to public health or safety.

Finally, the RTFA states that successful defendants *shall* receive attorney's fees and costs.

Just because there's an RTFA, however, doesn't prevent you from being sued. For instance, a cranberry farmer was sued by plaintiffs *including the state* for allegedly damaging a body of water by using phosphorous fertilizer. Even though the farmer offered proof that his farm was following best operating practices as per the state university's guidelines, the plaintiffs argued that use of the fertilizer represented a "substantial threat to public health or safety." The case devolved into an incredibly expensive, semantic debate over whether his use of phosphorous fertilizer constituted a "substantial threat" to public health.

With support from his insurer (thanks to the possibility of recovering fees and costs) the farmer had to fight all the way up to the state supreme court.[541] He prevailed and was awarded fees and costs as was the insurer.

---

[541] *State v. Zawistowski*, 309 Wis. 2d 233 (Wis. Ct. App. 2008) (unpublished opinion). The cranberry farmer's saga is documented in Attorney Tiffany Dowell's law review article, "Daddy Won't Sell the Farm: Drafting Right To Farm Statutes To Protect Small Family Producers," 18 S.J. Agric. L. Rev. 127, 142-43 (2009). Dowell is also the author of the Texas Agricultural Law Blog, agrilife.org/texasaglaw.

Speaking of insurance, on a somewhat related note, the Wisconsin Supreme Court has indicated that manure (the bacteria contained therein) is a pollutant that may not be covered if your insurance policy excludes pollution. Hence, an insurance company had no duty to defend a farmer from a suit alleging that they negligently spread liquid cow manure on their property (through part of an approved nutrient management plan) and thereby polluted their neighbors' wells.[542] You may want to make sure your insurance policy is up to date and realize that the RTFA may not protect you in case of pollution as injurious to public health.

As to zoning ordinances and regulations, an appellate court in Wisconsin held that nothing in the RTFA "prevents local governments from regulating agricultural uses and practices absent a finding that those uses or practices meet the heightened nuisance standard set forth in the statute."[543] Hence, the RTFA didn't prevent a town ordinance regulating propane cannons from being applied to a farmer, even though they agreed that the use of cannons is an "agricultural practice" and that he'd been using them for decades prior to the town's enactment of a "scare gun" ordinance. Granted, he was firing them every twenty-six seconds right next to the neighbor's residence and wasn't particularly sorry about it.

Neither was his use of propane cannons grandfathered in as a "prior non-conforming use" because his "use" of the land was for farming not for firing scare guns. He had a right to continue to use the land for farming, not to employ a particular farming practice said the court. Moreover, the court explained that the ordinance wasn't a zoning ordinance but was a regulatory ordinance enacted pursuant to the town's non-zoning police power. Neither was the scare gun permit "a land use permit" which

[542] *Wilson Mut. Ins. Co. v. Falk*, 857 N.W.2d 156 (Wis. 2014).
[543] *Town of Trempealeau v. Klein*, 870 N.W.2d 247 (Wis. Ct. App. 2015) (unpublished decision), available at cases.justia.com/wisconsin/court-of-appeals/2015-2014ap002719.pdf.

would have been restricted by the county's prohibition allowing general agricultural practices in all agricultural districts without issuance of a "land use permit." Finally, the legislative purpose statement in the RTFA that urges local governments to use their zoning authority to address conflicts between agriculture and other land uses, is practically worthless since it doesn't expressly withdraw a local government's authority to regulate agricultural practices using their non-zoning police power.[544]

The court concluded that the scare gun town ordinance was a good example of a town using its non-zoning police power to keep the peace between agricultural and non-agricultural use.

Another perspective might be that the Wisconsin RTFA needs a serious overhaul if it's to keep up with best practices in modern agriculture-oriented states as relates to protecting generally accepted agricultural practices not only from private nuisance actions but from folks using zoning boards to put farms out of business.

In the case of the propane cannon, a statute allowing generally accepted agricultural practices absent a substantial change would likely have handled the situation just fine. As would adding a one or two year statute of repose as many states have done. In fairness to the zoning board in the propane cannon case, it looks as though they did a lot of homework in framing their regulation; however, the legislative purpose, which amounted to a mere "hope" that zoning boards will respect agricultural uses and practices, was shown in that decision to worth no more than the paper it's printed on. (Again, even though the board in that case was fairly respectful, according to the court decision it didn't have to be.)

It's actually surprising that the law hasn't yet been amended based on the ruling in that case. Then again, maybe the farmer's activity and attitude didn't elicit much sympathy.

---

[544] Contrast this with the decision in Illinois, *Vill. of Lafayette v. Brown*, 27 N.E.3d 687 (Ill. App. 2015), appeal denied *Vill. of Lafayette v. Brown*, 392 Ill. Dec. 370 (Ill. 2015).

Wyoming: Wyoming Statutes § 11-44-103[545] & 104[546]

In 2015, the state legislature passed a bill (not a constitutional amendment) protecting the Right to Farm. "To protect agriculture as a vital part of the economy of Wyoming," said the legislature, "the rights of farmers and ranchers to engage in farm or ranch operations shall be forever guaranteed in this state."[547]

It's not immediately clear what effect such aspirational language will have on the application of Wyoming law. For example, no express language in the statute restricts zoning ordinances, though there may be some exemptions for agricultural uses or structures in other statutes. However, the language from the new Right to Farm act might have some effect on those provisions because Wyoming is a Dillon's Rule state, meaning localities can only act on direct authorization from the state.[548] It might be argued that ordinances that violate the Right to Farm statute are null and void.

Already, farm or ranch operations are given absolute immunity from public or private nuisance if they conform to "generally accepted agricultural management practices," if they existed prior to a change in the use of adjacent land, and if they wouldn't have been a nuisance before the change in land use that

---

[545] law.justia.com/codes/wyoming/2011/title11/chapter44/section11-44-103.

[546] Added recently. Bill available here: legisweb.state.wy.us/2015/bills/SF0009.pdf.

[547] Wisc. Stat. § 11-44-104 (2016). The statute goes on to say, "Nothing in this section shall be construed to modify any provision of common law or statutes relating to trespass, eminent domain, existing or previously enacted laws or rules or any other property rights."

[548] Although the Illinois RTFA does not contain language nullifying local zoning ordinances, an appellate court held that an ordinance banning all commercial farming in a village violated the purposes of the RTFA to protect and promote agricultural use as against nuisance and was therefore preempted. See the following case as well as the extensive discussion under the Illinois section of this survey, *Vill. of Lafayette v. Brown*, 27 N.E.3d 687 (Ill. App. 2015), appeal denied *Vill. of Lafayette v. Brown*, 392 Ill. Dec. 370 (Ill. 2015). There was also speculation that the failed constitutional amendment in Oklahoma may have affected contrary ordinances. K. Querry, "Battle over 'Right to Farm' proposal continues across Oklahoma" KFOR.com, 5:24 PM, July 24, 2016, kfor.com/2016/07/24/battle-over-right-to-farm-proposal-continues-across-oklahoma.

gave rise to the complaint.[549]

The statute doesn't seem to define "generally accepted agricultural management practices." That is the only place the
phrase appears in the code. You'll have to fight it out in court.

Farm or ranch operations, on the other hand, are defined as:

> [T]he science and art of production of plants and animals useful to
> man except those listed under [wildlife definitions], including, but
> not limited to, the preparation of these products for man's use and
> their disposal by marketing or otherwise, and includes horticulture,
> floriculture, viticulture, silviculture, dairy, livestock, poultry, bee and
> any and all forms of farm and ranch products and farm and ranch
> production.[550]

Until 1991, Right to Farm was limited to a nuisance exemption for cases brought against feedlots by neighbors whose
ownership was subsequent to the establishment of the feedlot
and provided the feedlot is in compliance with state regulations.[551] That exemption still exists alongside the RTFA.

Like several states, Wyoming has a statute on the books that
imposes penalties for gathering "resource data information"; it
basically criminalizes undercover recording operations.[552] That
statute was recently upheld as being constitutional in federal
district court.[553]

---

[549] Wyo. Stat. § 11-44-103 (2016).
[550] Wyo. Stat. § 11-44-102 (2016).
[551] Wyo. Stat. § 11-39-102 (2016).
[552] Wyo. Stat. Ann. § 6-3-414 (2016).
[553] *Western Watersheds Project v. Michael*, 2016 U.S. Dist. LEXIS 88843 at *28 (D. Wyo.
July 6, 2016)

www.ingramcontent.com/pod-product-compliance
Lightning Source LLC
Chambersburg PA
CBHW061254220326
41599CB00028B/5649